ENGINEERING CHEMISTRY
Basic Principles

ENGINEERING CHEMISTRY
Basic Principles

BY

CRA PAYNE
MATILDA MOORE

KNOWLEDGE
BAKERS

KNOWLEDGE BAKERS

ISBN: 978-93-9001-387-6 (HB)

Published by : **Knowledge Bakers**
Pune, India
Email: info@knowledgebakers.com

PREFACE

Engineering Chemistry stands at the crossroads of chemistry and engineering, serving as a foundational discipline that underpins advancements in various engineering fields, including materials science, environmental engineering, chemical engineering, and more. By elucidating the fundamental principles of chemistry and their applications in engineering contexts, this discipline equips students and practitioners with the knowledge and skills necessary to tackle complex challenges and drive innovation in diverse industries. As we embark on this journey through the basic principles of Engineering Chemistry, it becomes apparent that a strong foundation in fundamental concepts is essential for understanding the intricate interplay between chemistry and engineering and for fostering creative problem-solving in real-world scenarios.

This book, titled "Engineering Chemistry: Basic Principles," is designed to provide a comprehensive introduction to the core concepts and principles that form the basis of Engineering Chemistry. Through collaboration with experienced educators and practitioners in the field, we aim to offer students and readers a clear and concise overview of essential topics, from atomic structure and chemical bonding to thermodynamics, kinetics, and electrochemistry.

As editors of this volume, we have curated a collection of chapters that cover a wide range of fundamental principles in Engineering Chemistry, presenting each topic in a systematic and accessible manner. Each chapter is structured to provide a conceptual framework for understanding key principles, followed by illustrative examples, practical applications, and problem-solving exercises to reinforce learning and facilitate deeper comprehension.

Furthermore, this book seeks to highlight the interdisciplinary nature of Engineering Chemistry and its relevance to various engineering disciplines and real-world applications. By showcasing examples of how chemical principles are applied in engineering practice, from designing new materials to optimizing industrial processes and addressing environmental challenges, we aim to inspire readers to appreciate the interconnectedness of chemistry and engineering and to recognize the vast potential for innovation and impact in this field.

As we embark on this enlightening journey through the basic principles of Engineering Chemistry, we extend our gratitude to the contributors whose expertise and dedication have enriched this volume. We hope that "Engineering Chemistry: Basic Principles" serves as a valuable resource for students, educators, and practitioners alike, empowering them to develop a solid foundation in Engineering Chemistry and to embark on a journey of discovery and innovation in the dynamic intersection of chemistry and engineering.

Cra Payne
Matilda Moore

CONTENTS

	Preface	v
1.	Engineering Materials: Overview	1
2.	High Polymers for Engineering Applications	69
3.	Principles of Chromatography	93
4.	Classical Methods: Chemical Analysis	114
5.	Chemical Fuels	132
6.	Instrumental Methods of Analysis	146
7.	Gravimetric Analysis	166
8.	Introduction to Polymers and Polymerization	206
9.	Laws of Thermodynamics	217
	Bibliography	245
	Index	249

1

ENGINEERING MATERIALS: OVERVIEW

Engineering materials refers to the group of materials that are used in the construction of manmade structures and components. The primary function of an engineering material is to withstand applied loading without breaking and without exhibiting excessive deflection. The major classifications of engineering materials include metals, polymers, ceramics, and composites.

The chemical composition of engineering material indicates the elements which are combined together to form that material. Chemical composition of a material effects the properties of engineering materials very much. The strength, hardness, ductility, brittleness, corrosion resistance, weldability etc. depends on chemical composition of materials. Hence, we should also have the knowledge of chemical composition of engineering materials.

Atomic bonding represents how atoms are bounded to each other to form the material. Many properties, such as melting point, boiling point, thermal conductivity and electrical conductivity of materials are governed by atomic bonding of materials. Hence, to understand the properties of materials, it is very important to study the atomic bonding of materials.

CORROSION RESISTANCE

Corrosion is a gradual chemical or electromechemical attack on a metal by its surrounding medium. Due to the corrosion, metal starts getting converted into an oxide, salt or some other compound. Corrosion of a metals is effected by many factors such as air, industrial atmosphere, acid, bases, slat solutions and soils etc. Corrosion has a very adverse effect on materials. Due to corrosion, the strength and life of a material is reduced.

Corrosion resistance of a material is the ability of material to resist the oxidation in atmospheric condition. Generally pure metals such as iron, copper, aluminum etc. gets corroded in slowly in atmosphere. To avoid the corrosion of these metal in pure form, we use these metals in the form of alloys such as stainless steel, brass, bronze, German silver, Gunmetal etc.

Acidity or Alkalinity

Acidity or Alkalinity is an important chemical property of engineering materials. A material is acetic or Alkane, it is decided by the ph value of the material. Ph value of a material varies from 0 to 14. Ph value of 7 is considered to be neutral. Ordinary water is having ph value of 7. The materials which are having ph value below 7 are called Acetic and Materials which are having ph value greater than 7 are called alkane. Acidity of Alkalinity of material indicates that how they react with other materials.

New materials have been among the greatest achievements of every age and they have been central to the growth, prosperity, security, and quality of life of humans since the beginning of history. It is always new materials that open the door to new technologies, whether they are in civil, chemical, construction, nuclear, aeronautical, agricultural, mechanical, biomedical or electrical engineering.

Materials scientists and engineers continue to be at the forefront of all of these and many other areas of science, too. Materials science and engineering influences our lives each time we buy or use a new device, machine, or structure. (You can read more about the impact of this exciting field in our list of suggested readings.) The definition of the academic field of Materials Science & Engineering stems from a realization concerning every application of materials: it is the properties of the material that give it value. A material may be chosen for its strength, its electrical properties, resistance to heat or corrosion, or a host of other reasons; but they all relate to properties.

Experience shows that all of the useful properties of a material are intimately related to its structure, at all levels, including which atoms are present, how the atoms are joined, and how groups of atoms are arranged throughout the material. Most importantly, we learn how this structure, and the resulting properties, are controlled by the processing of the material.

Finally materials must perform their tasks in an economical and societally responsible manner. Understanding the relationships between properties, structure, processing and performance makes the Materials Engineer the master of the engineering universe.

Fundamentals

The basis of materials science involves relating the desired properties and relative performance of a material in a certain application to the structure of the atoms and phases in that material through characterization. The major determinants of the structure of a material and thus of its properties are its constituent chemical elements and the way in which it has been processed into its final form. These characteristics, taken together and related through the laws of thermodynamics, govern a material's microstructure, and thus its properties.

The manufacture of a perfect crystal of a material is currently physically impossible. Instead materials scientists manipulate the defects in crystalline materials such as precipitates, grain boundaries (Hall–Petch relationship), interstitial atoms, vacancies or substitutional atoms, to create materials with the desired properties.

Not all materials have a regular crystal structure. Polymers display varying degrees of crystallinity, and many are completely non-crystalline. Glasses, some ceramics, and many natural materials are amorphous, not possessing any long-range order in their atomic arrangements. The study of polymers combines elements of chemical and statistical thermodynamics to give thermodynamic, as well as mechanical, descriptions of physical properties.

In addition to industrial interest, materials science has gradually developed into a field which provides tests for condensed matter or solid state theories. New physics emerge because of the diverse new material properties which need to be explained.

Classes of Materials

Materials science encompasses various classes of materials, each of which may constitute a separate field. There are several ways to classify materials. For instance by the type of bonding between the atoms. The traditional groups are ceramics, metals and polymers based on atomic structure and chemical composition. New materials have resulted in more classes. One way of classifying materials is:

- Biomaterials
- Carbon
- Ceramics
- Composite materials
- Glass

- Metals
- Nanomaterials
- Polymers
- Refractory
- Semiconductors
- Thin Films
- Functionally Graded Materials

Materials in industry

Radical materials advances can drive the creation of new products or even new industries, but stable industries also employ materials scientists to make incremental improvements and troubleshoot issues with currently used materials. Industrial applications of materials science include materials design, cost-benefit tradeoffs in industrial production of materials, processing techniques (casting, rolling, welding, ion implantation, crystal growth, thin-film deposition, sintering, glassblowing, etc.), and analytical techniques (characterization techniques such as electron microscopy, x-ray diffraction, calorimetry, nuclear microscopy (HEFIB), Rutherford backscattering, neutron diffraction, small-angle X-ray scattering (SAXS), etc.).

Besides material characterization, the material scientist/engineer also deals with the extraction of materials and their conversion into useful forms. Thus ingot casting, foundry techniques, blast furnace extraction, and electrolytic extraction are all part of the required knowledge of a metallurgist/engineer. Often the presence, absence or variation of minute quantities of secondary elements and compounds in a bulk material will have a great impact on the final properties of the materials produced, for instance, steels are classified based on 1/10 and 1/100 weight percentages of the carbon and other alloying elements they contain. Thus, the extraction and purification techniques employed in the extraction of iron in the blast furnace will have an impact of the quality of steel that may be produced.

The overlap between physics and materials science has led to the offshoot field of materials physics, which is concerned with the physical properties of materials. The approach is generally more macroscopic and applied than in condensed matter physics. See important publications in materials physics for more details on this field of study.

Ceramics and glasses

Another application of the material sciences is the structures of glass and ceramics, typically associated with the most brittle materials. Bonding in ceramics and glasses are using covalent and ionic-covalent types with SiO2 (silica or sand) as a fundamental building block. Ceramics are as soft as clay and as hard as stone and concrete. Usually, they are crystalline in form. Most glasses contain a metal oxide fused with silica. At high temperatures used to prepare glass, the material is a viscous liquid. The structure of glass forms into an amorphous state upon cooling. Windowpanes and eyeglasses are important examples. Fibers of glass are also available. Diamond and carbon in its graphite form are considered to be ceramics.

Engineering ceramics are known for their stiffness and stability under high temperatures, compression and electrical stress. Alumina, silicon carbide, and tungsten carbide are made from a fine powder of their constituents in a process of sintering with a binder. Hot pressing provides higher density material. Chemical vapor deposition can place a film of a ceramic on another material. Cermets are ceramic particles containing some metals. The wear resistance of tools is derived from cemented carbides with the metal phase of cobalt and nickel typically added to modify properties.

Composite materials

Filaments are commonly used for reinforcement in composite materials.

Another application of material science in industry is the making of composite materials. Composite materials are structured materials composed of two or more macroscopic phases. Applications range from structural elements such as steel-reinforced concrete, to the thermally insulative tiles which play a key and integral role in NASA's Space Shuttle thermal protection system which is used to protect the surface of the shuttle from the heat of re-entry into the Earth's atmosphere. One example is reinforced Carbon-Carbon (RCC), the light gray material which withstands re-entry temperatures up to 1510 °C (2750 °F) and protects the Space Shuttle's wing leading edges and nose cap. RCC is a laminated composite material made from graphite rayon cloth and impregnated with a phenolic resin. After curing at high temperature in an autoclave, the laminate is pyrolized to convert the resin to carbon, impregnated with furfural alcohol in a vacuum chamber, and cured/pyrolized to convert the furfural alcohol to carbon. In order to provide oxidation resistance for reuse capability, the outer layers of the RCC are converted to silicon carbide.

Other examples can be seen in the "plastic" casings of television sets, cell-phones and so on. These plastic casings are usually a composite material made up of a thermoplastic matrix such as acrylonitrile-butadiene-styrene (ABS) in which calcium carbonate chalk, talc, glass fibers or carbon fibers have been added for added strength, bulk, or electrostatic dispersion. These additions may be referred to as reinforcing fibers, or dispersants, depending on their purpose.

Polymers

Polymers are also an important part of materials science. Polymers are the raw materials (the resins) used to make what we commonly call plastics. Plastics are really the final product, created after one or more polymers or additives have been added to a resin during processing, which is then shaped into a final form. Polymers which have been around, and which are in current widespread use, include polyethylene, polypropylene, PVC, polystyrene, nylons, polyesters, acrylics, polyurethanes, and polycarbonates. Plastics are generally classified as "commodity", "specialty" and "engineering" plastics.

PVC (polyvinyl-chloride) is widely used, inexpensive, and annual production quantities are large. It lends itself to an incredible array of applications, from artificial leather to electrical insulation and cabling, packaging and containers. Its fabrication and processing are simple and well-established. The versatility of PVC is due to the wide range of plasticisers and other additives that it accepts. The term "additives" in polymer science refers to the chemicals and compounds added to the polymer base to modify its material properties.

Polycarbonate would be normally considered an engineering plastic (other examples include PEEK, ABS). Engineering plastics are valued for their superior strengths and other special material properties. They are usually not used for disposable applications, unlike commodity plastics.

Specialty plastics are materials with unique characteristics, such as ultra-high strength, electrical conductivity, electro-fluorescence, high thermal stability, etc.

The dividing lines between the various types of plastics is not based on material but rather on their properties and applications. For instance, polyethylene (PE) is a cheap, low friction polymer commonly used to make disposable shopping bags and trash bags, and is considered a commodity plastic, whereas medium-density polyethylene (MDPE) is used for underground gas and water pipes, and another variety called Ultra-high Molecular Weight Polyethylene UHMWPE is an engineering plastic which is used extensively as the glide rails for industrial equipment and the low-friction socket in implanted hip joints.

Metals

Metals are the most commonly used class of engineering material. Metal alloys are especially common, and they are formed by combining a metal with one or more other metallic and/or non-metallic materials. The combination usually occurs through a process of melting, mixing, and cooling. The goal of alloying is to improve the properties of the base material in some desirable way.

Metal alloy compositions are described in terms of the percentages of the various elements in the alloy, where the percentages are measured by weight.

Ferrous Alloys

Ferrous alloys have iron as the base element. These alloys and include steels and cast irons. Ferrous alloys are the most common metal alloys in use due to the abundance of iron, ease of production, and high versatility of the material. The biggest disadvantage of many ferrous alloys is low corrosion resistance.

Carbon is an important alloying element in all ferrous alloys. In general, higher levels of carbon increase strength and hardness, and decrease ductility and weldability.

Carbon Steel

Carbon steels are basically just mixtures of iron and carbon. They may contain small amounts of other elements, but carbon is the primary alloying ingredient. The effect of adding carbon is an increase in strength and hardness.

Most carbon steels are plain carbon steels, of which there are several types.

Low-Carbon Steel

Low-carbon steel has less than about 0.30% carbon. It is characterized by low strength but high ductility. Some strengthening can be achieved through cold working, but it does not respond well to heat treatment. Low-carbon steel is very weldable and is inexpensive to produce. Common uses for low-carbon steel include wire, structural shapes, machine parts, and sheet metal.

Medium-Carbon Steel

Medium-carbon steel contains between about 0.30% to 0.70% carbon. It can be heat treated to increase strength, especially with the higher carbon contents. Medium-carbon steel is frequently used for axles, gears, shafts, and machine parts.

High-Carbon Steel

High-carbon steel contains between about 0.70% to 1.40% carbon. It has high strength but low ductility. Common uses include drills, cutting tools, knives, and springs.

Biomaterial

A biomaterial is any matter, surface, or construct that interacts with biological systems. As a science, biomaterials is about fifty years old. The study of biomaterials is called biomaterials science. It has experienced steady and strong growth over its history, with many companies investing large amounts of money into the development of new products. Biomaterials science encompasses elements of medicine, biology, chemistry, tissue engineering and materials science.

Biomaterials can be derived either from nature or synthesized in the laboratory using a variety of chemical approaches utilizing metallic components or ceramics. They are often used and/or adapted for a medical application, and thus comprises whole or part of a living structure or biomedical device which performs, augments, or replaces a natural function. Such functions may be benign, like being used for a heart valve, or may be bioactive with a more interactive functionality such as hydroxy-apatite coated hip implants. Biomaterials are also used every day in dental applications, surgery, and drug delivery. For example, a construct with impregnated pharmaceutical products can be placed into the body, which permits the prolonged release of a drug over an extended period of time. A biomaterial may also be an autograft, allograft or xenograft used as a transplant material.

Materials scientists are currently paying more and more attention to the process inorganic crystallization within a largely organic matrix of naturally occurring compounds. This process typically occurs at ambient temperature and pressure. Interestingly, the vital organisms through which these crystalline minerals form are capable of consistently producing intricately complex structures. Understanding the processes in which living organisms are capable of regulating the growth of crystalline minerals such as silica could lead to significant scientific advances and novel synthesis techniques for nanoscale composite materials—or nanocomposites.

Biomineralization is the process by which living organisms produce minerals, often to harden or stiffen existing tissues. Such tissues are called mineralized tissues. It is an extremely widespread phenomenon; all six taxonomic kingdoms contain members that are able to form minerals, and over 60 different minerals have been identified in organisms. Examples include silicates in algae and diatoms, carbonates in invertebrates, and calcium phosphates and carbonates in vertebrates. These minerals often form structural features such as sea shells and the bone in mammals and birds. Organisms have been producing mineralised skeletons for the past 550 million years. Other examples include copper, iron and gold deposits involving bacteria. Biologically-formed minerals often have special uses such as magnetic sensors in magnetotactic bacteria (Fe3O4), gravity sensing devices (CaCO3, CaSO4, BaSO4) and iron storage and mobilization (Fe2O3•H2O in the protein ferritin).

In terms of taxonomic distribution, the most common biominerals are the phosphate and carbonate salts of calcium that are used in conjunction with organic polymers such as collagen and chitin to give structural support to bones and shells. The structures of these biocomposite materials are highly controlled from the nanometer to the macroscopic level, resulting in complex architectures that provide multifunctional properties. Because this range of control over mineral growth is desirable for materials engineering applications, there is significant interest in understanding and elucidating the mechanisms of biologically controlled biomineralization.

Ceramic

A ceramic is an inorganic, nonmetallic solid prepared by the action of heat and subsequent cooling. Ceramic materials may have a crystalline or partly crystalline structure, or may be amorphous (e.g., a glass). Because most common ceramics are crystalline, the definition of ceramic is often restricted to inorganic crystalline materials, as opposed to the noncrystalline glasses.

The earliest ceramics were pottery objects or 27,000 year old figurines made from clay, either by itself or mixed with other materials, hardened in fire. Later ceramics were glazed and fired to create a colored, smooth surface. Ceramics now include domestic, industrial and building products and art objects. In the 20th century, new ceramic materials were developed for use in advanced ceramic engineering; for example, in semiconductors.

The word "ceramic" comes from the Greek word êåñáìéêüò (keramikos), "of pottery" or "for pottery", from êÝñáìïò (keramos), "potter's clay, tile, pottery". The earliest mention of the root "ceram-" is the Mycenaean Greek ke-ra-me-we, "workers of ceramics", written in Linear B syllabic script. "Ceramic" may be used as an adjective describing a material, product or process; or as a singular noun, or, more commonly, as a plural noun, "ceramics".

Types of ceramic products

For convenience, ceramic products are usually divided into four sectors; these are shown below with some examples:

- Structural, including bricks, pipes, floor and roof tiles
- Refractories, such as kiln linings, gas fire radiants, steel and glass making crucibles
- Whitewares, including tableware, cookware, wall tiles, pottery products and sanitary ware
- Technical, is also known as engineering, advanced, special, and in Japan, fine ceramics. Such items include tiles used in the Space Shuttle program, gas burner nozzles, ballistic protection, nuclear fuel uranium oxide pellets, biomedical implants, coatings of jet engine turbine blades, ceramic disk brake, missile nose cones, bearing (mechanical). Frequently, the raw materials do not include clays.

Examples of whiteware ceramics

- Earthenware, which is often made from clay, quartz and feldspar.
- Stoneware
- Porcelain, which is often made from kaolin
- Bone china

Classification of technical ceramics

Technical ceramics can also be classified into three distinct material categories:

- Oxides: alumina, beryllia, ceria, zirconia
- Nonoxides: carbide, boride, nitride, silicide
- Composite materials: particulate reinforced, fiber reinforced, combinations of oxides and nonoxides.

Each one of these classes can develop unique material properties because ceramics tend to be crystalline.

Other applications of ceramics

- Knife blades: the blade of a ceramic knife will stay sharp for much longer than that of a steel knife, although it is more brittle and can be snapped by dropping it on a hard surface.
- Vehicle ceramic brake discs are resistant to abrasion at high temperatures.

- Advanced composite ceramic and metal matrices have been designed for most modern armoured fighting vehicles because they offer superior penetrating resistance against shaped charges such as high explosive antitank (HEAT) rounds and kinetic energy penetrators.
- Ceramics such as alumina and boron carbide have been used in ballistic armored vests to repel large-caliber rifle fire. Such plates are known commonly as small arms protective inserts (SAPIs). Similar material is used to protect cockpits of some military airplanes, because of the low weight of the material.
- Ceramic balls can be used to replace steel in ball bearings. Their higher hardness means they are much less susceptible to wear and can offer more than triple lifetimes. They also deform less under load, meaning they have less contact with the bearing retainer walls and can roll faster. In very high speed applications, heat from friction during rolling can cause problems for metal bearings, which are reduced by the use of ceramics. Ceramics are also more chemically resistant and can be used in wet environments where steel bearings would rust. In some cases, their electricity-insulating properties may also be valuable in bearings. The two major drawbacks to using ceramics are a significantly higher cost and susceptibility to damage under shock loads.
- In the early 1980s, Toyota researched production of an adiabatic engine using ceramic components in the hot gas area. The ceramics would have allowed temperatures of over 3000°F (1650°C). The expected advantages would have lighter materials, no or reduced cooling system, and hence a major weight reduction. The expected increase of fuel efficiency of the engine (caused by the higher temperature, as shown by Carnot's theorem) could not be verified experimentally; it was found that the heat transfer on the hot ceramic cylinder walls is higher than the transfer to a cooler metal wall. Obviously the cooler gas film on the metal surface works as a thermal insulator. Thus, despite all of these desirable properties, such engines have not succeeded in production because of costs for the ceramic components and the limited advantages. (Small imperfections in the ceramic material with its low fracture toughness lead to cracks, which can lead to potentially dangerous equipment failure.) Such engines are possible in laboratory settings, but mass production is not feasible with current technology.
- Work is being done in developing ceramic parts for gas turbine engines. Currently, even blades made of advanced metal alloys used in the engines'

hot section require cooling and careful limiting of operating temperatures. Turbine engines made with ceramics could operate more efficiently, giving aircraft greater range and payload for a set amount of fuel.

- Recent advances have been made in ceramics which include bioceramics, such as dental implants and synthetic bones. Hydroxyapatite, the natural mineral component of bone, has been made synthetically from a number of biological and chemical sources and can be formed into ceramic materials. Orthopedic implants coated with these materials bond readily to bone and other tissues in the body without rejection or inflammatory reactions so are of great interest for gene delivery and tissue engineering scaffolds. Most hydroxyapatite ceramics are very porous and lack mechanical strength, and are used to coat metal orthopedic devices to aid in forming a bond to bone or as bone fillers. They are also used as fillers for orthopedic plastic screws to aid in reducing the inflammation and increase absorption of these plastic materials. Work is being done to make strong, fully dense nano crystalline hydroxyapatite ceramic materials for orthopedic weight bearing devices, replacing foreign metal and plastic orthopedic materials with a synthetic, but naturally occurring, bone mineral. Ultimately, these ceramic materials may be used as bone replacements or with the incorporation of protein collagens, synthetic bones.
- High-tech ceramic is used in watchmaking for producing watch cases. The material is valued by watchmakers for its light weight, scratch resistance, durability and smooth touch. IWC is one of the brands that initiated the use of ceramic in watchmaking. The case of the IWC 2007 Top Gun edition of the Pilot's Watch double chronograph is crafted in black ceramic.

TYPES OF CERAMIC MATERIALS

A ceramic material is an inorganic, non-metallic, often crystalline oxide, nitride or carbide material. Some elements, such as carbon or silicon, may be considered ceramics. Ceramic materials are brittle, hard, strong in compression, weak in shearing and tension. They withstand chemical erosion that occurs in other materials subjected to acidic or caustic environments. Ceramics generally can withstand very high temperatures, such as temperatures that range from 1,000 °C to 1,600 °C (1,800 °F to 3,000 °F). A glass is often not understood as a ceramic because of its amorphous (noncrystalline) character. However, glass making involves several steps of the ceramic process and its mechanical properties are similar to ceramic materials.

Traditional ceramic raw materials include clay minerals such as kaolinite, whereas more recent materials include aluminium oxide, more commonly known as alumina. The modern ceramic materials, which are classified as advanced ceramics, include silicon carbide and tungsten carbide. Both are valued for their abrasion resistance, and hence find use in applications such as the wear plates of crushing equipment in mining operations. Advanced ceramics are also used in the medicine, electrical and electronics industries.

Crystalline ceramics

Crystalline ceramic materials are not amenable to a great range of processing. Methods for dealing with them tend to fall into one of two categories – either make the ceramic in the desired shape, by reaction in situ, or by "forming" powders into the desired shape, and then sintering to form a solid body. Ceramic forming techniques include shaping by hand (sometimes including a rotation process called "throwing"), slip casting, tape casting (used for making very thin ceramic capacitors, e.g.), injection molding, dry pressing, and other variations. Details of these processes are described in the two books listed below.[which?] A few methods use a hybrid between the two approaches.

Noncrystalline ceramics

Noncrystalline ceramics, being glasses, tend to be formed from melts. The glass is shaped when either fully molten, by casting, or when in a state of toffee-like viscosity, by methods such as blowing to a mold. If later heat treatments cause this glass to become partly crystalline, the resulting material is known as a glass-ceramic, widely used as cooktop and also as a glass composite material for nuclear waste disposal.

Ceramics in archaeology

Ceramic artifacts have an important role in archaeology for understanding the culture, technology and behavior of peoples of the past. They are among the most common artifacts to be found at an archaeological site, generally in the form of small fragments of broken pottery called sherds. Processing of collected sherds can be consistent with two main types of analysis: technical and traditional.

Traditional analysis involves sorting ceramic artifacts, sherds and larger fragments into specific types based on style, composition, manufacturing and morphology. By creating these typologies it is possible to distinguish between different cultural styles, the purpose of the ceramic and technological state of the people among other conclusions. In addition, by looking at stylistic changes of ceramics over time is it possible to separate (seriate) the ceramics into distinct diagnostic groups (assemblages). A comparison of ceramic artifacts with known dated assemblages allows for a chronological assignment of these pieces.

The technical approach to ceramic analysis involves a finer examination of the composition of ceramic artifacts and sherds to determine the source of the material and through this the possible manufacturing site. Key criteria are the composition of the clay and the temper used in the manufacture of the article under study: temper is a material added to the clay during the initial production stage, and it is used to aid the subsequent drying process. Types of temper include shell pieces, granite fragments and ground sherd pieces called 'grog'. Temper is usually identified by microscopic examination of the temper material. Clay identification is determined by a process of refiring the ceramic, and assigning a color to it using Munsell Soil Color notation. By estimating both the clay and temper compositions, and locating a region where both are known to occur, an assignment of the material source can be made. From the source assignment of the artifact further investigations can be made into the site of manufacture.

Ceramic engineering

Ceramic engineering is the science and technology of creating objects from inorganic, non-metallic materials. This is done either by the action of heat, or at lower temperatures using precipitation reactions from high purity chemical solutions. The term includes the purification of raw materials, the study and production of the chemical compounds concerned, their formation into components and the study of their structure, composition and properties.

Ceramic materials may have a crystalline or partly crystalline structure, with long-range order on atomic scale. Glass ceramics may have an amorphous or glassy structure, with limited or short-range atomic order. They are either formed from a molten mass that solidifies on cooling, formed and matured by the action of heat, or chemically synthesized at low temperatures using, for example, hydrothermal or sol-gel synthesis.

The special character of ceramic materials gives rise to many applications in materials engineering, electrical engineering, chemical engineering and mechanical engineering. As ceramics are heat resistant, they can be used for many tasks that materials like metal and polymers are unsuitable for. Ceramic materials are used in a wide range of industries, including mining, aerospace, medicine, refinery, food and chemical industries, packaging science, electronics, industrial and transmission electricity, and guided lightwave transmission.

Modern industry

Now a multi-billion dollar a year industry, ceramic engineering and research has established itself as an important field of science. Applications continue to expand as researchers develop new kinds of ceramics to serve different purposes.

- Zirconium dioxide ceramics are used in the manufacture of knives. The blade of the ceramic knife will stay sharp for much longer than that of a steel knife, although it is more brittle and can be snapped by dropping it on a hard surface.
- Ceramics such as alumina, boron carbide and silicon carbide have been used in bulletproof vests to repel large-caliber rifle fire. Such plates are known commonly as small-arms protective inserts (SAPI). Similar material is used to protect cockpits of some military airplanes, because of the low weight of the material.
- Silicon nitride parts are used in ceramic ball bearings. Their higher hardness means that they are much less susceptible to wear and can offer more than triple lifetimes. They also deform less under load meaning they have less contact with the bearing retainer walls and can roll faster. In very high speed applications, heat from friction during rolling can cause problems for metal bearings; problems which are reduced by the use of ceramics. Ceramics are also more chemically resistant and can be used in wet environments where steel bearings would rust. The major drawback to using ceramics is a significantly higher cost. In many cases their electrically insulating properties may also be valuable in bearings.
- In the early 1980s, Toyota researched production of an adiabatic ceramic engine which can run at a temperature of over 6000 °F (3300 °C). Ceramic engines do not require a cooling system and hence allow a major weight reduction and therefore greater fuel efficiency. Fuel efficiency of the engine is also higher at high temperature, as shown by Carnot's theorem. In a conventional metallic engine, much of the energy released from the fuel must be dissipated as waste heat in order to prevent a meltdown of the metallic parts. Despite all of these desirable properties, such engines are not in production because the manufacturing of ceramic parts in the requisite precision and durability is difficult. Imperfection in the ceramic leads to cracks, which can lead to potentially dangerous equipment failure. Such engines are possible in laboratory settings, but mass-production is not feasible with current technology.

- Work is being done in developing ceramic parts for gas turbine engines. Currently, even blades made of advanced metal alloys used in the engines' hot section require cooling and careful limiting of operating temperatures. Turbine engines made with ceramics could operate more efficiently, giving aircraft greater range and payload for a set amount of fuel.
- Recently, there have been advances in ceramics which include bio-ceramics, such as dental implants and synthetic bones. Hydroxyapatite, the natural mineral component of bone, has been made synthetically from a number of biological and chemical sources and can be formed into ceramic materials. Orthopedic implants made from these materials bond readily to bone and other tissues in the body without rejection or inflammatory reactions. Because of this, they are of great interest for gene delivery and tissue engineering scaffolds. Most hydroxyapatite ceramics are very porous and lack mechanical strength and are used to coat metal orthopedic devices to aid in forming a bond to bone or as bone fillers. They are also used as fillers for orthopedic plastic screws to aid in reducing the inflammation and increase absorption of these plastic materials. Work is being done to make strong, fully dense nano crystalline hydroxyapatite ceramic materials for orthopedic weight bearing devices, replacing foreign metal and plastic orthopedic materials with a synthetic, but naturally occurring, bone mineral. Ultimately these ceramic materials may be used as bone replacements or with the incorporation of protein collagens, synthetic bones.
- High-tech ceramic is used in watchmaking for producing watch cases. The material is valued by watchmakers for its light weight, scratch-resistance, durability and smooth touch. IWC is one of the brands that initiated the use of ceramic in watchmaking. The case of the IWC 2007 Top Gun edition of the Pilot's Watch Double chronograph is crafted in high-tech black ceramic.

Glass-ceramics

Glass-ceramic materials share many properties with both glasses and ceramics. Glass-ceramics have an amorphous phase and one or more crystalline phases and are produced by a so-called "controlled crystallization", which is typically avoided in glass manufacturing. Glass-ceramics often contain a crystalline phase which constitutes anywhere from 30% [m/m] to 90% [m/m] of its composition by volume, yielding an array of materials with interesting thermomechanical properties.

In the processing of glass-ceramics, molten glass is cooled down gradually before reheating and annealing. In this heat treatment the glass partly crystallizes. In many cases, so-called 'nucleation agents' are added in order to regulate and control the crystallization process. Because there is usually no pressing and sintering, glass-ceramics do not contain the volume fraction of porosity typically present in sintered ceramics.

The term mainly refers to a mix of lithium and aluminosilicates which yields an array of materials with interesting thermomechanical properties. The most commercially important of these have the distinction of being impervious to thermal shock. Thus, glass-ceramics have become extremely useful for countertop cooking. The negative thermal expansion coefficient (TEC) of the crystalline ceramic phase can be balanced with the positive TEC of the glassy phase. At a certain point (~70% crystalline) the glass-ceramic has a net TEC near zero. This type of glass-ceramic exhibits excellent mechanical properties and can sustain repeated and quick temperature changes up to 1000 °C.

Processing Steps

The traditional ceramic process generally follows this sequence: Milling '! Batching '! Mixing '! Forming '! Drying '! Firing '! Assembly.

- Milling is the process by which materials are reduced from a large size to a smaller size. Milling may involve breaking up cemented material (in which case individual particles retain their shape) or pulverization (which involves grinding the particles themselves to a smaller size). Milling is generally done by mechanical means, including attrition (which is particle-to-particle collision that results in agglomerate break up or particle shearing), compression (which applies a forces that results in fracturing), and impact (which employs a milling medium or the particles themselves to cause fracturing). Attrition milling equipment includes the wet scrubber (also called the planetary mill or wet attrition mill), which has paddles in water creating vortexes in which the material collides and break up. Compression mills include the jaw crusher, roller crusher and cone crusher. Impact mills include the ball mill, which has media that tumble and fracture the material. Shaft impactors cause particle-to particle attrition and compression.
- Batching is the process of weighing the oxides according to recipes, and preparing them for mixing and drying.
- Mixing occurs after batching and is performed with various machines, such as dry mixing ribbon mixers (a type of cement mixer), Mueller mixers, and pug mills. Wet mixing generally involves the same equipment.

- Forming is making the mixed material into shapes, ranging from toilet bowls to spark plug insulators. Forming can involve: (1) Extrusion, such as extruding "slugs" to make bricks, (2) Pressing to make shaped parts, (3) Slip casting, as in making toilet bowls, wash basins and ornamentals like ceramic statues. Forming produces a "green" part, ready for drying. Green parts are soft, pliable, and over time will lose shape. Handling the green product will change its shape. For example, a green brick can be "squeezed", and after squeezing it will stay that way.
- Drying is removing the water or binder from the formed material. Spray drying is widely used to prepare powder for pressing operations. Other dryers are tunnel dryers and periodic dryers. Controlled heat is applied in this two-stage process. First, heat removes water. This step needs careful control, as rapid heating causes cracks and surface defects. The dried part is smaller than the green part, and is brittle, necessitating careful handling, since a small impact will cause crumbling and breaking.
- Firing is where the dried parts pass through a controlled heating process, and the oxides are chemically changed to cause sintering and bonding. The fired part will be smaller than the dried part.

Forming Methods

Ceramic forming techniques include throwing, slipcasting, tape casting, injection molding, dry pressing, isostatic pressing, hot isostatic pressing (HIP) and others. Methods for forming ceramic powders into complex shapes are desirable in many areas of technology. Such methods are required for producing advanced, high-temperature structural parts such as heat engine components and turbines. Materials other than ceramics which are used in these processes may include: wood, metal, water, plaster and epoxy—most of which will be eliminated upon firing.

These forming techniques are well known for providing tools and other components with dimensional stability, surface quality, high (near theoretical) density and microstructural uniformity. The increasing use and diversity of specialty forms of ceramics adds to the diversity of process technologies to be used.

Thus, reinforcing fibers and filaments are mainly made by polymer, sol-gel, or CVD processes, but melt processing also has applicability. The most widely used specialty form is layered structures, with tape casting for electronic substrates and packages being preeminent. Photolithography is of increasing interest for precise patterning of conductors and other components for such packaging. Tape casting or forming processes are also of increasing interest for other applications, ranging from open structures such as fuel cells to ceramic composites. The other major layer structure is coating, where melt spraying is very important, but chemical and physical vapor deposition and chemical (e.g., sol-gel and polymer pyrolysis) methods are all seeing increased use. Besides open structures from formed tape, extruded structures, such as honeycomb catalyst supports, and highly porous structures, including various foams, for example, reticulated foam, are of increasing use.

Densification of consolidated powder bodies continues to be achieved predominantly by (pressureless) sintering. However, the use of pressure sintering by hot pressing is increasing, especially for non-oxides and parts of simple shapes where higher quality (mainly microstructural homogeneity) is needed, and larger size or multiple parts per pressing can be an advantage.

Ceramic composites

Substantial interest has arisen in recent years in fabricating ceramic composites. While there is considerable interest in composites with one or more non-ceramic constituents, the greatest attention is on composites in which all constituents are ceramic. These typically comprise two ceramic constituents: a continuous matrix, and a dispersed phase of ceramic particles, whiskers, or short (chopped) or continuous ceramic fibers. The challenge, as in wet chemical processing, is to obtain a uniform or homogeneous distribution of the dispersed particle or fiber phase.

Consider first the processing of particulate composites. The particulate phase of greatest interest is tetragonal zirconia because of the toughening that can be achieved from the phase transformation from the metastable tetragonal to the monoclinic crystalline phase, aka transformation toughening. There is also substantial interest in dispersion of hard, non-oxide phases such as SiC, TiB, TiC, boron, carbon and especially oxide matrices like alumina and mullite. There is also interest too incorporating other ceramic particulates, especially those of highly anisotropic thermal expansion. Examples include Al2O3, TiO2, graphite, and boron nitride.

In processing particulate composites, the issue is not only homogeneity of the size and spatial distribution of the dispersed and matrix phases, but also control of the matrix grain size. However, there is some built-in self-control due to inhibition of matrix grain growth by the dispersed phase. Particulate composites, though generally offer increased resistance to damage, failure, or both, are still quite sensitive to inhomogeneities of composition as well as other processing defects such as pores. Thus they need good processing to be effective.

Particulate composites have been made on a commercial basis by simply mixing powders of the two constituents. Although this approach is inherently limited in the homogeneity that can be achieved, it is the most readily adaptable for existing ceramic production technology. However, other approaches are of interest.

Composite Material

Composite materials (also called composition materials or shortened to composites) are materials made from two or more constituent materials with significantly different physical or chemical properties, that when combined, produce a material with characteristics different from the individual components. The individual components remain separate and distinct within the finished structure.

Typical engineered composite materials include:

- Composite building materials such as cements, concrete
- Reinforced plastics such as fiber-reinforced polymer
- Metal Composites
- Ceramic Composites (composite ceramic and metal matrices)

Composite materials are generally used for buildings, bridges and structures such as boat hulls, race car bodies, shower stalls, bathtubs, and storage tanks, imitation granite and cultured marble sinks and countertops. The most advanced examples perform routinely on spacecraft in demanding environments.

Materials

Concrete is the most common artificial composite material of all and typically consists of loose stones (aggregate) held with a matrix of cement. Concrete is a very robust material, much more robust than cement, and will not compress or shatter even under quite a large compressive force. However, concrete cannot survive tensile loading (i.e. if stretched it will quickly break apart). Therefore to give concrete the ability to resist being stretched, steel bars, which can resist high stretching forces, are often added to concrete to form reinforced concrete.

Fibre-reinforced polymers or FRPs include carbon-fibre reinforced plastic or CFRP, and glass-reinforced plastic or GRP. If classified by matrix then there are thermoplastic composites, short fibre thermoplastics, long fibre thermoplastics or long fibre-reinforced thermoplastics. There are numerous thermoset composites, but advanced systems usually incorporate aramid fibre and carbon fibre in an epoxy resin matrix.

Shape memory polymer composites are high-performance composites, formulated using fibre or fabric reinforcement and shape memory polymer resin as the matrix. Since a shape memory polymer resin is used as the matrix, these composites have the ability to be easily manipulated into various configurations when they are heated above their activation temperatures and will exhibit high strength and stiffness at lower temperatures. They can also be reheated and reshaped repeatedly without losing their material properties. These composites are ideal for applications such as lightweight, rigid, deployable structures; rapid manufacturing; and dynamic reinforcement.

Composites can also use metal fibres reinforcing other metals, as in metal matrix composites (MMC) or ceramic matrix composites (CMC), which includes bone (hydroxyapatite reinforced with collagen fibres), cermet (ceramic and metal) and concrete. Ceramic matrix composites are built primarily for fracture toughness, not for strength. Organic matrix/ceramic aggregate composites include asphalt concrete, mastic asphalt, mastic roller hybrid, dental composite, syntactic foam and mother of pearl. Chobham armour is a special type of composite armour used in military applications.

Additionally, thermoplastic composite materials can be formulated with specific metal powders resulting in materials with a density range from 2 g/cm^3 to 11 g/cm^3 (same density as lead). The most common name for this type of material is "high gravity compound" (HGC), although "lead replacement" is also used. These materials can be used in place of traditional materials such as aluminium, stainless steel, brass, bronze, copper, lead, and even tungsten in weighting, balancing (for example, modifying the centre of gravity of a tennis racquet), vibration damping, and radiation shielding applications. High density composites are an economically viable option when certain materials are deemed hazardous and are banned (such as lead) or when secondary operations costs (such as machining, finishing, or coating) are a factor.

A sandwich-structured composite is a special class of composite material that is fabricated by attaching two thin but stiff skins to a lightweight but thick core. The core material is normally low strength material, but its higher thickness provides the sandwich composite with high bending stiffness with overall low density.

Wood is a naturally occurring composite comprising cellulose fibres in a lignin and hemicellulose matrix. Engineered wood includes a wide variety of different products such as wood fibre board, plywood, oriented strand board, wood plastic composite (recycled wood fibre in polyethylene matrix), Pykrete (sawdust in ice matrix), Plastic-impregnated or laminated paper or textiles, Arborite, Formica (plastic) and Micarta. Other engineered laminate composites, such as Mallite, use a central core of end grain balsa wood, bonded to surface skins of light alloy or GRP. These generate low-weight, high rigidity materials.

Products

Fiber-reinforced composite materials have gained popularity (despite their generally high cost) in high-performance products that need to be lightweight, yet strong enough to take harsh loading conditions such as aerospace components (tails, wings, fuselages, propellers), boat and scull hulls, bicycle frames and racing car bodies. Other uses include fishing rods, storage tanks, and baseball bats. The new Boeing 787 structure including the wings and fuselage is composed largely of composites. Composite materials are also becoming more common in the realm of orthopedic surgery.

Carbon composite is a key material in today's launch vehicles and heat shields for the re-entry phase of spacecraft. It is widely used in solar panel substrates, antenna reflectors and yokes of spacecraft. It is also used in payload adapters, inter-stage structures and heat shields of launch vehicles. Furthermore disk brake systems of airplanes and racing cars are using carbon/carbon material, and the composite material with carbon fibers and silicon carbide matrix has been introduced in luxury vehicles and sports cars.

In 2007, an all-composite military Humvee was introduced by TPI Composites Inc and Armor Holdings Inc, the first all-composite military vehicle. By using composites the vehicle is lighter, allowing higher payloads. In 2008, carbon fiber and DuPont Kevlar (five times stronger than steel) were combined with enhanced thermoset resins to make military transit cases by ECS Composites creating 30-percent lighter cases with high strength.

Pipes and fittings for various purpose like transportation of potable water, fire-fighting, irrigation, seawater, desalinated water, chemical and industrial waste, and sewage are now manufactured in glass reinforced plastics.

Overview

Composites are made up of individual materials referred to as constituent materials. There are two main categories of constituent materials: matrix and reinforcement. At least one portion of each type is required. The matrix material surrounds and supports the reinforcement materials by maintaining their relative positions. The reinforcements impart their special mechanical and physical properties to enhance the matrix properties. A synergism produces material properties unavailable from the individual constituent materials, while the wide variety of matrix and strengthening materials allows the designer of the product or structure to choose an optimum combination.

Engineered composite materials must be formed to shape. The matrix material can be introduced to the reinforcement before or after the reinforcement material is placed into the mould cavity or onto the mould surface. The matrix material experiences a melding event, after which the part shape is essentially set. Depending upon the nature of the matrix material, this melding event can occur in various ways such as chemical polymerization or solidification from the melted state.

A variety of moulding methods can be used according to the end-item design requirements. The principal factors impacting the methodology are the natures of the chosen matrix and reinforcement materials. Another important factor is the gross quantity of material to be produced. Large quantities can be used to justify high capital expenditures for rapid and automated manufacturing technology. Small production quantities are accommodated with lower capital expenditures but higher labour and tooling costs at a correspondingly slower rate.

Many commercially produced composites use a polymer matrix material often called a resin solution. There are many different polymers available depending upon the starting raw ingredients. There are several broad categories, each with numerous variations. The most common are known as polyester, vinyl ester, epoxy, phenolic, polyimide, polyamide, polypropylene, PEEK, and others. The reinforcement materials are often fibres but also commonly ground minerals. The various methods described below have been developed to reduce the resin content of the final product, or the fibre content is increased. As a rule of thumb, lay up results in a product containing 60% resin and 40% fibre, whereas vacuum infusion gives a final product with 40% resin and 60% fiber content. The strength of the product is greatly dependent on this ratio.

Martin Hubbe and Lucian A Lucia consider Wood to be a natural composite of cellulose fibres in a matrix of lignin.

CONSTITUENTS

Matrices

Resins

Typically, most common polymer-based composite materials, including fiberglass, carbon fiber, and Kevlar, include at least two parts, the substrate and the resin.

Polyester resin tends to have yellowish tint, and is suitable for most backyard projects. Its weaknesses are that it is UV sensitive and can tend to degrade over time, and thus generally is also coated to help preserve it. It is often used in the making of surfboards and for marine applications. Its hardener is a peroxide, often MEKP (methyl ethyl ketone peroxide). When the peroxide is mixed with the resin, it decomposes to generate free radicals, which initiate the curing reaction. Hardeners in these systems are commonly called catalysts, but since they do not re-appear unchanged at the end of the reaction, they do not fit the strictest chemical definition of a catalyst. Vinylester resin tends to have a purplish to bluish to greenish tint. This resin has lower viscosity than polyester resin, and is more transparent. This resin is often billed as being fuel resistant, but will melt in contact with gasoline. This resin tends to be more resistant over time to degradation than polyester resin, and is more flexible. It uses the same hardeners as polyester resin (at a similar mix ratio) and the cost is approximately the same.

Epoxy resin is almost totally transparent when cured. In the aerospace industry, epoxy is used as a structural matrix material or as a structural glue.

Shape memory polymer (SMP) resins have varying visual characteristics depending on their formulation. These resins may be epoxy-based, which can be used for auto body and outdoor equipment repairs; cyanate-ester-based, which are used in space applications; and acrylate-based, which can be used in very cold temperature applications, such as for sensors that indicate whether perishable goods have warmed above a certain maximum temperature. These resins are unique in that their shape can be repeatedly changed by heating above their glass transition temperature (Tg). When heated, they become flexible and elastic, allowing for easy configuration. Once they are cooled, they will maintain their new shape. The resins will return to their original shapes when they are reheated above their Tg. The advantage of shape memory polymer resins is that they can be shaped and reshaped repeatedly without losing their material properties. These resins can be used in fabricating shape memory composites.

Other Matrices

Common matrices include mud (wattle and daub), cement (concrete), polymers (fiber reinforced plastics), metals and ceramics. Road surfaces are often made from asphalt concrete which uses bitumen as a matrix. Unusual matrices such as ice are sometime proposed as in pykecrete.

Reinforcement

Fiber

Reinforcement usually adds rigidity and greatly impedes crack propagation. Thin fibers can have very high strength, and provided they are mechanically well attached to the matrix they can greatly improve the composite's overall properties.

Fiber-reinforced composite materials can be divided into two main categories normally referred to as short fiber-reinforced materials and continuous fiber-reinforced materials. Continuous reinforced materials will often constitute a layered or laminated structure. The woven and continuous fibre styles are typically available in a variety of forms, being pre-impregnated with the given matrix (resin), dry, uni-directional tapes of various widths, plain weave, harness satins, braided, and stitched.

The short and long fibers are typically employed in compression moulding and sheet moulding operations. These come in the form of flakes, chips, and random mate (which can also be made from a continuous fibre laid in random fashion until the desired thickness of the ply / laminate is achieved).

Common fibers used for reinforcement include glass fibers, carbon fibers, cellulose (wood/paper fiber and straw) and high strength polymers for example aramid.

Other Reinforcment

Concrete uses aggregate, and reinforced concrete additionally uses steel bars (rebar) to tension the concrete. Steel mesh or wires are also used in some glass and plastic products.

Cores

Many composite layup designs also include a co-curing or post-curing of the prepreg with various other mediums, such as honeycomb or foam. This is commonly called a sandwich structure. This is a more common layup process for the manufacture of radomes, doors, cowlings, or non-structural parts.

Open- and closed-cell-structured foams like polyvinylchloride, polyurethane, polyethylene or polystyrene foams, balsa wood, syntactic foams, and honeycombs are commonly used core materials. Open- and closed-cell metal foam can also be used as core materials.

Glass

Glass is an amorphous (non-crystalline) solid material. Glasses are typically brittle and optically transparent.

The most familiar type of glass, used for centuries in windows and drinking vessels, is soda-lime glass, composed of about 75% silica (SiO2) plus sodium oxide

(Na_2O) from soda ash, lime (CaO), and several minor additives. Often, the term glass is used in a restricted sense to refer to this specific use.

In science, however, the term glass is usually defined in a much wider sense, including every solid that possesses a non-crystalline (i.e., amorphous) structure and that exhibits a glass transition when heated towards the liquid state. In this wider sense, glasses can be made of quite different classes of materials: metallic alloys, ionic melts, aqueous solutions, molecular liquids, and polymers. For many applications (bottles, eyewear) polymer glasses (acrylic glass, polycarbonate, polyethylene terephthalate) are a lighter alternative to traditional silica glasses.

GLASS INGREDIENTS

While fused quartz (primarily composed of SiO_2) is used for some special applications, it is not very common due to its high glass transition temperature of over 1200 °C (2192 °F). Normally, other substances are added to simplify processing. One is sodium carbonate (Na_2CO_3, "soda"), which lowers the glass transition temperature. However, the soda makes the glass water soluble, which is usually undesirable, so lime (calcium oxide [CaO], generally obtained from limestone), some magnesium oxide (MgO) and aluminium oxide (Al_2O_3) are added to provide for a better chemical durability. The resulting glass contains about 70 to 74% silica by weight and is called a soda-lime glass. Soda-lime glasses account for about 90% of manufactured glass.

Most common glass contains other ingredients added to change its properties. Lead glass or flint glass is more 'brilliant' because the increased refractive index causes noticeably more specular reflection and increased optical dispersion. Adding barium also increases the refractive index. Thorium oxide gives glass a high refractive index and low dispersion and was formerly used in producing high-quality lenses, but due to its radioactivity has been replaced by lanthanum oxide in modern eye glasses. Iron can be incorporated into glass to absorb infrared energy, for example in heat absorbing filters for movie projectors, while cerium(IV) oxide can be used for glass that absorbs UV wavelengths.

The following is a list of the more common types of silicate glasses, and their ingredients, properties, and applications:

1. Fused silica glass, vitreous silica glass: silica (SiO_2). Has very low thermal expansion, is very hard and resists high temperatures (1000–1500 °C). It is also the most resistant against weathering (alkali ions leaching out of the glass, while staining it). It is used for high temperature applications such as furnace tubes, melting crucibles, etc.

2. Soda-lime-silica glass, window glass: silica 72% + sodium oxide (Na2O) 14.2% + magnesia (MgO) 2.5% + lime (CaO) 10.0% + alumina (Al2O3) 0.6%. Is transparent, easily formed and most suitable for window glass. It has a high thermal expansion and poor resistance to heat (500–600 °C). Used for windows, containers, light bulbs, tableware.
3. Sodium borosilicate glass, Pyrex: silica 81% + boric oxide (B2O3) 12% + soda (Na2O) 4.5% + alumina (Al2O3) 2.0%. Stands heat expansion much better than window glass. Used for chemical glassware, cooking glass, car head lamps, etc. Borosilicate glasses (e.g. Pyrex) have as main constituents silica and boron oxide. They have fairly low coefficients of thermal expansion (7740 Pyrex CTE is 3.25×10–6/°C as compared to about 9×10"6/°C for a typical soda-lime glass), making them more dimensionally stable. The lower CTE also makes them less subject to stress caused by thermal expansion, thus less vulnerable to cracking from thermal shock. They are commonly used for reagent bottles, optical components and household cookware.
4. Lead-oxide glass, crystal glass: silica 59% + soda (Na2O) 2.0% + lead oxide (PbO) 25% + potassium oxide (K2O) 12% + alumina 0.4% + zinc oxide (ZnO) 1.5%. Has a high refractive index, making the look of glassware more brilliant (crystal glass). It also has a high elasticity, making glassware 'ring'. It is also more workable in the factory, but cannot stand heating very well.
5. Aluminosilicate glass: silica 57% + alumina 16% + boric oxide (B2O3) 4.0% + barium oxide (BaO) 6.0% + magnesia 7.0% + lime 10%. Extensively used for fiberglass, used for making glass-reinforced plastics (boats, fishing rods, etc.). Also for halogen bulb glass.
6. Oxide glass: alumina 90% + germanium oxide (GeO2) 10%. Extremely clear glass, used for fiber-optic wave guides in communication networks. Light loses only 5% of its intensity through 1 km of glass fiber.

Another common glass ingredient is "cullet" (recycled glass). The recycled glass saves on raw materials and energy; however, impurities in the cullet can lead to product and equipment failure.

Fining agents such as sodium sulfate, sodium chloride, or antimony oxide may be added to reduce the number of air bubbles in the glass mixture. Glass batch calculation is the method by which the correct raw material mixture is determined to achieve the desired glass composition.

Cement and Binding Material

Cement is a fine powder, which when mixed with water undergoes chemical change and thereafter allowed to set and harden is capable of uniting fragments or masses of solid matter together to produce a mechanically strong material. Cement can be used as binding material with water, for bonding solid particles of different sizes like bricks, stones or aggregate to form a monolith. Cements used in construction of buildings and civil engineering works contain compounds of lime, silica and alumina as their principal constituents and can be called as complex compounds.

Materials utilized in construction for making concretes and mortars, in bonding together individual elements of structural components, in waterproofing, and so on.

Mineral cementing materials. Mineral binding materials are powdered substances that have the capacity when mixed (with water) to form a plastic mass that later hardens into a strong, rocklike solid. Depending on their composition, basic properties, and field of utilization, mineral binding materials are subdivided into hydraulic, air-entrained, and acid-resistant types.

Hydraulic binding materials, after mixing with water and preliminary hardening in the air, continue to preserve and increase their strength in water; they may be used in underground and underwater hydraulic-engineering and other installations. Hydraulic binding materials include various cements: Portland cement and its varieties (quick-setting, plasticized, hydrophobic, plugging, sulfate-resistant, and nonstaining), pozzolanic cements (Portland-pozzolanic cement and limestone-pozzolanic cement), blast-furance cements (metallurgical cement, limestone-slag cement, and sulfate-slag cement), alumina and expanding cements, Roman cement, and hydraulic lime. In order to intensify the hardening processes in certain hydraulic binding materials (limestone-siliceous, limestone-slag, limestone-nephelinic, and others), steam treatment is utilized in autoclaves at a pressure ranging from 0.9 to 1.6 meganewtons per sq m (9 to 16 kilograms-force per sq cm) for a period of nine to ten hours.

Air-entrained binding materials can harden and maintain their strength after mixing only in the air; they are used solely for erecting aboveground structures that are not subject to the action of water. This group of materials includes gypsum cements (construction plaster, supersulfate cement, and highly fired gypsum), magnesian cements (caustic magnesite and dolomite), and air-hardening lime.

Acid-resistant binding materials can maintain their strength after hardening in air while in contact with acids; they are utilized in the construction of acid-resistant coverings. Such binding materials include acid-resistant cements, which are produced on the basis of water glass (sodium silicate) and acid-resistant microfillers and hardening accelerators. Various additives are introduced into binding materials in order to improve certain of their properties.

Organic binding materials. Organic binding materials are substances of organic origin (asphalt, bitumens, pitch, polyvinyl acetate, and furan resins) that have the capacity to make the transition from a plastic state to the hardness or slightly plastic states under the influence of physical or chemical processes.

Importance of cement in construction

Today, cement is a very important building material used in the construction industry. It was in 1824 that cement, as we know it today was developed. Joseph Aspdin of England first patented the manufacture of a very improvised type of cementitious material for building construction in 1824. The hardened cement paste resembled the natural stone occurring at Portland in England. Hence it was named as Portland cement.

Cement has replaced all other binding materials, for example, clay and lime, which ruled high for hundreds of years, in construction from last century. The reason being that, cement is the strongest binding material amongst clay, lime and cement.

Cement is now used in many construction trade items like Plain Concrete, Reinforced Concrete, Mortar, Plaster, Grouts, Paints and in Pre-cast Elements. In an ordinary building construction the cement accounts to about 15-17% of the cost of construction. In some other major constructions like concrete roads and bridges, it is very much more.

Some of the reasons for its popularity and universal acceptance are listed below:

1. Cement can be produced in large volumes in controlled condition, packed and transported
2. Cement is several times stronger binding material than lime and clay
3. It can be mixed and used at will with locally available materials at site
4. When stored properly in ordinary atmosphere does not deteriorate for reasonably longer time
5. When mixed with water, starts setting and acquires sufficient strength in a day or two, where as other binding materials require much longer time
6. When water is added to quick lime, lot of heat is generated, but in case of cement, heat generated is unnoticeable and comparatively much lesser

7. It can withstand compressive stresses well. Where tension and shear stresses occur it gives good bond to steel reinforcement and transfers excess stresses to steel
8. It is produced from the materials like limestone, hematite, bauxite, clay, etc which are abundantly available in the upper crust of the earth

LIME

Lime is a very fine white powder, used in mixes for mortar, plaster and render. It is made from limestone or chalk which is burnt in a kiln and becomes quicklime.

The quicklime is usually passed through a machine called a hydrator, where it combines with water and becomes hydrated lime. This is dried, crushed to a fine powder, then bagged and sold. Below is a diagram of the whole process.

Sometimes the lime is sold as quicklime, and the builder adds the water to it himself. This process is called "slaking" the lime or "running it to putty", and it is not described here.

Slaked lime and hydrated lime are chemically the same, but slaked lime has more water in it.

Hydraulic lime is made from limestone or chalk containing clay. It hardens when combined with water; and it also hardens well in damp places or even under water. It is stronger than other lime*;, although weaker than Portland cement.

Non-hydraulic lime comes from the purest limestones and chalks. It hardens by drying out and then slowly combining with the carbon dioxide in the air.

PORTLAND CEMENT

Cement is an adhesive substances of all kinds, but, in a narrower sense, the binding materials used in building and civil engineering construction. Cements of this kind are finely ground powders that, when mixed with water, set to a hard mass. Setting and hardening result from hydration, which is a chemical combination of the cement compounds with water that yields submicroscopic crystals or a gel-like material with a high surface area. Because of their hydrating properties, constructional cements, which will even set and harden under water, are often called hydraulic cements. The most important of these is portland cement.

Portland cement is a fine grey powder. Among the various kinds of cements, it is the most commonly used as binding material. It is made of a mixture of chalk or limestone and clay.

The following description of the manufacturing process is illustrated below.

The limestone or chalk and the clay, in appropriate proportions, are fed into a "wet grinding mill" and reduced to a creamy substance known as slurry. The slurry is pumped to a large cylindrical "kiln" which is about 90 m long and 3 m in diameter. The slurry enters the kiln at its upper end while pulverized (crushed) coal, gas or other fuel is blown in at the other end.

The temperature inside the kiln at the lower end is very intense, approximately 1500 degrees C; gradually decreasing towards the top end. So the slurry as it moves down the kiln is first dried, then heated, and then finally burnt. It leaves the kiln in the form of very hard "clinkers" shaped like small balls and of a dark brown to black in color. The clinkers are ground up to an extremely fine grey powder, which is the cement. The cement is packed in paper bags of 50 kg capacity.

Composition

Portland cement consists essentially of compounds of lime (calcium oxide, CaO) mixed with silica (silicon dioxide, SiO2) and alumina (aluminum oxide, Al2O3). The lime is obtained from a calcareous (lime-containing) raw material, and the other oxides are derived from an argillaceous (clayey) material. Additional raw materials such as silica sand, iron oxide (Fe2O3), and bauxite (containing hydrated aluminum, Al[OH]3) may be used in smaller quantities to get the desired composition.

The commonest calcareous raw materials are limestone and chalk, but others, such as coral or shell deposits, also are used. Clays, shales, slates, and estuarine muds are the common argillaceous raw materials. Marl, a compact calcareous clay, and cement rock contain both the calcareous and argillaceous components in proportions that sometimes approximate cement compositions. Another raw material is blast-furnace slag, which consists mainly of lime, silica, and alumina and is mixed with a calcareous material of high lime content. Kaolin, a white clay that contains little iron oxide, is used as the argillaceous component for white portland cement. Industrial wastes, such as fly ash and calcium carbonate from chemical manufacture, are other possible raw materials, but their use is small compared with that of the natural materials.

The magnesia (magnesium oxide, MgO) content of raw materials must be low because the permissible limit in portland cement is 4 to 5 percent. Other impurities in raw materials that must be strictly limited are fluorine compounds, phosphates, metal oxides and sulfides, and excessive alkalies.

Another essential raw material is gypsum, some 5 percent of which is added to the burned cement clinker during grinding to control the setting time of the cement. Portland cement also can be made in a combined process with sulfuric acid using calcium sulfate or anhydrite in place of calcium carbonate. The sulfur dioxide produced in the flue gases on burning is converted to sulfuric acid by normal processes.

Gypsum Plaster

Gypsum is a light weight material which is formed as the result of evaporating sea water in massive prehistoric basins. In terms of its chemical composition, it is Calcium Sulphate Di-hydrate (CaSO4.2H2O) and inherently contains 21% water by molecular weight.

Gypsum products have been used for centuries in the construction space and are the material of choice because of gypsum's unique properties. It has excellent insulation properties both thermal and acoustic and can offer very good passive fire protection, moisture resistance, impact resistance and vapour control when used in combination with the right systems.

Gypsum plaster, white cementing material made by partial or complete dehydration of the mineral gypsum, commonly with special retarders or hardeners added. Applied in a plastic state (with water), it sets and hardens by chemical recombination of the gypsum with water.

For especially hard finish plaster, the gypsum is completely dehydrated at high temperature, and such chemicals as alkali sulfate, alum, or borax are added. Hair or fibre and lime or clay may be added to the plasters during manufacture. The plaster coats, except for some finish coats, are sanded.

Gypsum plaster, or plaster of Paris, is produced by heating gypsum to about 300°F (150 °C):

2CaSO4·4H2O + Heat '! 2CaSO4·H2O + 3H2O (released as steam).

When the dry plaster powder is mixed with water, it re-forms into gypsum. The setting of unmodified plaster starts about 10 minutes after mixing and is complete in about 45 minutes; but not fully set for 72 hours. If plaster or gypsum is heated above 392°F (200°C), anhydrite is formed, which will also re-form as gypsum if mixed with water.

A large gypsum deposit at Montmartre in Paris led gypsum plaster to be commonly known as "plaster of Paris".

Plasterers often use gypsum to simulate the appearance of surfaces of wood, stone, or metal, on movie and theatrical sets for example. Nowadays, theatrical plasterers often use expanded polystyrene, although the job title remains unchanged.

Plaster of Paris can be used to impregnate gauze bandages to make a sculpting material called modroc. It is used similarly to clay, as it is easily shaped when wet, yet sets into a resilient and lightweight structure. This is the material which was (and sometimes still is) used to make classic plaster orthopedic casts to protect limbs with broken bones. Set modroc is an early example of a composite material.

Lime plaster

Lime plaster is a mixture of calcium hydroxide and sand (or other inert fillers). Carbon dioxide in the atmosphere causes the plaster to set by transforming the calcium hydroxide into calcium carbonate (limestone). Whitewash is based on the same chemistry. Lime plaster is type of plaster composed of hydrated lime, sand and water. Lime plaster is similar to Lime mortar, the main difference is the based on use rather than composition. Traditional lime plaster contains also horse hair to reinforce plaster.

It is sold as 'bagged' powder or hydrated lime; or is available as lime putty. Lime putty is generally considered to be more suitable for pure lime application.

Non-hydraulic lime is the most commonly used and known lime, also called (high) calcium lime or air lime, as it sets only by reaction with CO2 in the air and will not set until dry. This causes limitations in construction use as the lime can remain soft for months or years. Non-hydraulic lime can only set through carbonatation (re-absorption of CO2).

Hydraulic lime and hydrated lime must not be confused. Hydrated lime is merely a form in which lime can be supplied (as opposed to quicklime or lime putty); while 'hydraulic' refers to its ability to set under water, or in wet conditions.

To make lime plaster, limestone (calcium carbonate) is heated to produce quicklime (calcium oxide). Water is then added to produce slaked lime (calcium hydroxide), which is sold as a wet putty or a white powder. Additional water is added to form a paste prior to use. The paste may be stored in air-tight containers. Once exposed to the atmosphere, the calcium hydroxide turns back into calcium carbonate, causing the plaster to set.

Lime plaster was a common building material for wall surfaces in a process known as lath and plaster, whereby a series of wooden strips on a studwork frame was covered with a semi-dry plaster that hardened into a surface. The plaster used in most lath and plaster construction was mainly lime plaster, with a cure time of about a month. To stabilize the lime plaster during curing, small amounts of plaster of Paris were incorporated into the mix. Because plaster of Paris sets quickly, "retardants" were used to slow setting time enough to allow workers to mix large working quantities of lime putty plaster. A modern form of this method uses expanded metal mesh over wood or metal structures, which allows a great freedom of design as it is adaptable to both simple and compound curves. Today this building method has been partly replaced with drywall, also composed mostly of gypsum plaster. In both these methods a primary advantage of the material is that it is resistant to a fire within a room and so can assist in reducing or eliminating structural damage or destruction provided the fire is promptly extinguished.

Lime plaster is used for frescoes, where pigments, diluted in water, are applied to the still wet plaster. USA and Iran are the main plaster producers in the world.

Lime is an extremely caustic material when wet, with a pH of 12. (Lime when carbonated can be up a ph of 8.6). As such, the use of protective goggles, gloves, and clothing are necessary when working with lime. Clean water should also be kept readily accessible for first aid purposes when working with lime in case of accidental eye or skin exposure.

- First aid for cases of skin exposure to lime involves neutralization with very mild acid such as vinegar or lemon juice.
- First aid for cases of accidental eye exposure consists of repeatedly flushing the eye for several minutes with fresh water. Medical attention should be sought in such cases.

Cement plaster

Cement plaster is a mixture of suitable plaster, sand, portland cement and water which is normally applied to masonry interiors and exteriors to achieve a smooth surface. Interior surfaces sometimes receive a final layer of gypsum plaster. Walls constructed with stock bricks are normally plastered while face brick walls are not plastered. Various cement-based plasters are also used as proprietary spray fireproofing products. These usually use vermiculite as lightweight aggregate. Heavy versions of such plasters are also in use for exterior fireproofing, to protect LPG vessels, pipe bridges and vessel skirts.

Use

In recent years, the construction sector has witnessed a number of new trends, technology advancements and innovations across applications, all aimed at making construction faster and delivering higher performance.

However, very few would have imagined that the substitute for the generations old method of sand cement surface plastering will be an even older method of plastering i.e. gypsum plastering. Gypsum has proved to be a miraculous material aiding interior construction. Gypsum surface plaster is one form that has caught the attention on the sector and is growing in popularity.

The oldest example of usage of gypsum as an internal plaster is found in the Pyramids of Giza in Egypt which were lined with gypsum stucco plaster centuries ago and are still intact, which is proof of the durability & performance of this product.

When a wall is constructed with brick or block, they are finished with a coat of plaster which is called internal plastering. Internal plastering has traditionally been done through a process of sand-cement plastering finished with POP pruning.

This is a two stage process and involves multiple elements like sand, cement and water which is mixed onsite. This form of plastering is being slowly replaced by a direct single coat application of gypsum plaster.

So gypsum plaster is actually a replacement of two processes i.e. sand cement plastering & POP punning. Gypsum plaster is produced as a result of calcinations of the raw gypsum in an automatic kettle under controlled temperature.

Uses of Gypsum in Buildings

Gypsum building materials are used in all construction types (residential, non-residential, new or refurbished), ranging from complex high-tech systems to easy to install products adapted for use by the general public.

More than 1,600 million m2 of European interior surfaces are covered every year with plasterboards every year. around 5 million tons of plaster are used in Europe for interior lining.

More than 1,500 million m2 of European interior surfaces are covered every year with plasterboards, blocks or plaste

A. Plasterboard

Plasterboard is used for partitions and the lining of walls, ceilings, roofs and floors. The properties of plasterboard can be modified to meet specific requirements, such as fire resistance, humidity resistance, shock resistance, etc.

B Decorative Plaster

Plaster powder, mixed with water, manually or through the use of silo-supplied spray systems, are used to create an effective and aesthetically-pleasing lining for brick and block walls, and for ceilings.

Gypsum's adaptability in application lends itself to moulding and shaping. Since time immemorial, Gypsum has been used by skilled craftsmen to create decorative plaster mouldings.

C. Building plaster

Gypsum plaster is used for walls and ceilings.

D. Plaster blocks

Gypsum blocks are used for partitions and Gypsum tiles for ceilings.

E. Gypsum-based self-levelling screeds

Anhydrite or Alpha-Hemihydrates are used in the production of self-levelling floor screeds.

F. Gypsum Fibreboards

Gypsum fibreboard is used for partitions and the lining of walls, ceilings, roofs and floors. Standard Gypsum fibreboard offers good performance when it comes to shock resistance, sound insulation and humidity resistance.

REFRACTORY

A refractory material is one that retains its strength at high temperatures. ASTM C71 defines refractories as "non-metallic materials having those chemical and physical properties that make them applicable for structures, or as components of systems, that are exposed to environments above 1,000 °F (811 K; 538 °C)".

Refractory materials are used in linings for furnaces, kilns, incinerators and reactors. They are also used to make crucibles. Refractory materials must be chemically and physically stable at high temperatures. Depending on the operating environment, they need to be resistant to thermal shock, be chemically inert, and/or have specific ranges of thermal conductivity and of the coefficient of thermal expansion.

The oxides of aluminium (alumina), silicon (silica) and magnesium (magnesia) are the most important materials used in the manufacturing of refractories. Another oxide usually found in refractories is the oxide of calcium (lime). Fire clays are also widely used in the manufacture of refractories.

Refractories must be chosen according to the conditions they will face. Some applications require special refractory materials. Zirconia is used when the material must withstand extremely high temperatures. Silicon carbide and carbon (graphite) are two other refractory materials used in some very severe temperature conditions, but they cannot be used in contact with oxygen, as they will oxidize and burn.

Binary compounds such as tungsten carbide or boron nitride can be very refractory. Hafnium carbide is the most refractory binary compound known, with a melting point of 3890 °C. The ternary compound tantalum hafnium carbide has one of the highest melting points of all known compounds (4215 °C).

Classification of refractory materials

Refractories can be classified on the basis of chemical composition, method of manufacture, physical form or according to their applications.

Based on chemical composition

Acidic refractories

They consist of mostly acidic materials like alumina (Al2O3) and silica (SiO2). They are not attacked or affected by acidic materials, but easily affected by basic materials e.g.:-silica, alumina, fire clay refractories, etc.

Neutral refractories

These are used in areas where slags and atmosphere are either acidic or basic and are chemically stable to both acids and bases. The main raw materials belongs to, but not confined to, R2O3 group. The common examples of these materials are alumina (Al2O3), chromia (Cr2O3) and carbon.

Basic refractories

These are used on areas where slags and atmosphere are basic; they are stable to alkaline materials but react with acids. The main raw materials belong to the RO group to which magnesia (MgO) is a very common example. Other examples include dolomite and chrome-magnesia.

Based on method of manufacture

1. Dry press process
2. Fused cast
3. Hand molded
4. Formed (normal, fired or chemically bonded)
5. Un-formed (monolithic-plastic, ramming and gunning mass, castables)
6. Un-formed Dry Vibratable refractories.

SHAPED

These have fixed size and shapes. These may be further divided into standard shapes and special shapes. Standard shapes have dimension that are conformed by most refractory manufacturers and are generally applicable to kilns or furnaces of the same types. Special shapes are specifically made for particular kilns or furnaces.

Unshaped (Monolithic refractories)

These are without definite form and are only given shape upon application. These types are better known as monolithic refractories. The common examples are plastic masses, Ramming masses, castables, gunning masses, fettling mix, mortars etc.

Dry vibration linings often used in Induction furnace linings are also monolithic, and sold and transported as a dry powder, usually with a magnesia/alumina composition with additions of other chemicals for altering specific properties. They are also finding more applications in blast furnace linings, although this use is still rare.

Nanomaterials

Nanomaterials is a field that takes a materials science-based approach to nanotechnology. It studies materials with morphological features on the nanoscale, and especially those that have special properties stemming from their nanoscale dimensions. Nanoscale is usually defined as smaller than a one tenth of a micrometer in at least one dimension, though this term is sometimes used for even smaller materials.

On 18 October 2011, the European Commission adopted the following definition of a nanomaterial:

A natural, incidental or manufactured material containing particles, in an unbound state or as an aggregate or as an agglomerate and where, for 50% or more of the particles in the number size distribution, one or more external dimensions is in the size range 1 nm – 100 nm. In specific cases and where warranted by concerns for the environment, health, safety or competitiveness the number size distribution threshold of 50% may be replaced by a threshold between 1 and 50%.

An important aspect of nanotechnology is the vastly increased ratio of surface area to volume present in many nanoscale materials, which makes possible new quantum mechanical effects. One example is the "quantum size effect" where the electronic properties of solids are altered with great reductions in particle size. Nanoparticles, for example, take advantage of their dramatically increased surface area to volume ratio. Their optical properties, e.g. fluorescence, become a function of the particle diameter. This effect does not come into play by going from macro to micro dimensions. However, it becomes pronounced when the nanometer size range is reached.

A certain number of physical properties also alter with the change from macroscopic systems. Novel mechanical properties of nanomaterials is a subject of nanomechanics research. When brought into a bulk material, nanoparticles can strongly influence the mechanical properties of the material, like stiffness or elasticity. For example, traditional polymers can be reinforced by nanoparticles resulting in novel materials which can be used as lightweight replacements for metals. Such nanotechnologically enhanced materials may enable a weight reduction accompanied by an increase in stability and improved functionality. Catalytic activities also reveal new behaviour in the interaction with biomaterials.

Clay

Clay is a general term including many combinations of one or more clay minerals with traces of metal oxides and organic matter. Geologic clay deposits are mostly composed of phyllosilicate minerals containing variable amounts of water trapped in the mineral structure. Clay minerals are typically formed over long periods of time by the gradual chemical weathering of rocks, usually silicate-bearing, by low concentrations of carbonic acid and other diluted solvents. These solvents, usually acidic, migrate through the weathering rock after leaching through upper weathered layers. In addition to the weathering process, some clay minerals are formed by hydrothermal activity. Clay deposits may be formed in place as residual deposits in soil, but thick deposits usually are formed as the result of a secondary sedimentary deposition process after they have been eroded and transported from their original location of formation. Clay deposits are typically associated with very low energy depositional environments such as large lakes and marine basins.

Primary clays, also known as kaolins, are located at the site of formation. Secondary clay deposits have been moved by erosion and water from their primary location.

lays exhibit plasticity when mixed with water in certain proportions. When dry, clay becomes firm and when fired in a kiln, permanent physical and chemical changes occur. These reactions, among other changes, cause the clay to be converted into a ceramic material. Because of these properties, clay is used for making pottery items, both utilitarian and decorative. Different types of clay, when used with different minerals and firing conditions, are used to produce earthenware, stoneware, and porcelain. Prehistoric humans discovered the useful properties of clay. Some of the earliest pottery shards recovered are from central Honshu, Japan. They are associated with the Jomon culture and the deposits from which they were recovered have been radiocarbon dated to around 14000BC. Depending on the content of the soil, clay can appear in various colors, from a dull gray to a deep orange-red.

Clay tablets were used as the first known writing medium, inscribed with cuneiform script through the use of a blunt reed called a stylus. Purpose-made clay balls were also used as sling ammunition.

Clays sintered in fire were the first form of ceramic. Bricks, cooking pots, art objects, dishware, and even musical instruments such as the ocarina can all be shaped from clay before being fired. Clay is also used in many industrial processes, such as paper making, cement production, and chemical filtering. Clay is also often used in the manufacture of pipes for smoking tobacco. Until the late 20th century bentonite clay was widely used as a mold binder in the manufacture of sand castings.

Clay, being relatively impermeable to water, is also used where natural seals are needed, such as in the cores of dams, or as a barrier in landfills against toxic seepage (lining the landfill, preferably in combination with geotextiles).

Rocket propellant

Rocket propellant is mass ejected as exhaust to create thrust for a rocket. There are several types of rocket propellant currently in use. Along with the most commonly used propellants, scientists are continuing to look for and develop new and more efficient kinds.

Rocket propellant is a material used by a rocket as, or to produce in a chemical reaction, the reaction mass (propulsive mass) that is ejected, typically with very high speed, from a rocket engine to produce thrust, and thus provide spacecraft propulsion.

In a chemical rocket propellants undergo exothermic chemical reactions to produce hot gas. There may be a single propellant, or multiple propellants; in the latter case one can distinguish fuel and oxidizer. The gases produced expand and push on a nozzle, which accelerates them until they rush out of the back of the rocket at extremely high speed.

For smaller attitude control thrusters, a compressed gas escapes the spacecraft through a propelling nozzle. A potential other method is that the propellant is not burned but just heated. In ion propulsion, the propellant is made of electrically charged atoms (ions), which are electromagnetically pushed out of the back of the spacecraft. Magnetically accelerated ion drives are not usually considered to be rockets however, but a similar class of thrusters use electrical heating and magnetic nozzles. In pulse propulsion, a heavy, metallic base acquires the force from an explosion behind it, for example from an atomic bomb, and a transfers it to a dampening system that reduces the shock to the payload.

First there are the propellants for solid rockets. Most of us are already familiar with gun powder. However there are others that are commonly used in space flight. The booster rockets used in space shuttle launches are solid rockets. In their case, the solid propellant is mostly ammonium perchlorate, mixed with aluminum, iron oxide, and a polymer as a binding agent to hold the mixture together. The solid fuel in hybrid rockets is paraffin, a substance that is being highly touted because its exhaust – carbon dioxide and water – is environmentally friendly. Other types include "candy" propellant, which uses a sugar-based fuel and composite which is normally a mixture like the one mentioned above. The mixtures vary to get different types of thrust. Project Orion, a proposed nuclear pulse rocket used plastic as its propellant as it would break down under nuclear fission into lighter hydrogen atoms.

Propellant is the chemical mixture burned to produce thrust in rockets and consists of a fuel and an oxidizer. A fuel is a substance that burns when combined with oxygen producing gas for propulsion. An oxidizer is an agent that releases oxygen for combination with a fuel. The ratio of oxidizer to fuel is called the mixture ratio. Propellants are classified according to their state - liquid, solid, or hybrid.

The gauge for rating the efficiency of rocket propellants is specific impulse, stated in seconds. Specific impulse indicates how many pounds (or kilograms) of thrust are obtained by the consumption of one pound (or kilogram) of propellant in one second. Specific impulse is characteristic of the type of propellant, however, its exact value will vary to some extent with the operating conditions and design of the rocket engine.

Liquid Propellants

In a liquid propellant rocket, the fuel and oxidizer are stored in separate tanks, and are fed through a system of pipes, valves, and turbopumps to a combustion chamber where they are combined and burned to produce thrust. Liquid propellant engines are more complex than their solid propellant counterparts, however, they offer several advantages. By controlling the flow of propellant to the combustion chamber, the engine can be throttled, stopped, or restarted.

A good liquid propellant is one with a high specific impulse or, stated another way, one with a high speed of exhaust gas ejection. This implies a high combustion temperature and exhaust gases with small molecular weights. However, there is another important factor that must be taken into consideration: the density of the propellant. Using low-density propellants means that larger storage tanks will be required, thus increasing the mass of the launch vehicle. Storage temperature is also important. A propellant with a low storage temperature, i.e. a cryogenic, will require thermal insulation, thus further increasing the mass of the launcher. The toxicity of the propellant is likewise important. Safety hazards exist when handling, transporting, and storing highly toxic compounds. Also, some propellants are very corrosive; however, materials that are resistant to certain propellants have been identified for use in rocket construction.

Liquid propellants used in rocketry can be classified into three types: petroleum, cryogens, and hypergols.

Petroleum fuels are those refined from crude oil and are a mixture of complex hydrocarbons, i.e. organic compounds containing only carbon and hydrogen. The petroleum used as rocket fuel is a type of highly refined kerosene, called RP-1 in the United States. Petroleum fuels are usually used in combination with liquid oxygen as the oxidizer. Kerosene delivers a specific impulse considerably less than cryogenic fuels, but it is generally better than hypergolic propellants.

Specifications for RP-1 where first issued in the United States in 1957 when the need for a clean burning petroleum rocket fuel was recognized. Prior experimentation with jet fuels produced tarry residue in the engine cooling passages and excessive soot, coke and other deposits in the gas generator. Even with the new specifications, kerosene-burning engines still produce enough residues that their operational lifetimes are limited.

Liquid oxygen and RP-1 are used as the propellant in the first-stage boosters of the Atlas and Delta II launch vehicles. It also powered the first-stages of the Saturn 1B and Saturn V rockets.

Cryogenic propellants are liquefied gases stored at very low temperatures, most frequently liquid hydrogen (LH2) as the fuel and liquid oxygen (LO2 or LOX) as the oxidizer. Hydrogen remains liquid at temperatures of -253 oC (-423 oF) and oxygen remains in a liquid state at temperatures of -183 oC (-297 oF).

Because of the low temperatures of cryogenic propellants, they are difficult to store over long periods of time. For this reason, they are less desirable for use in military rockets that must be kept launch ready for months at a time. Furthermore, liquid hydrogen has a very low density (0.071 g/ml) and, therefore, requires a storage volume many times greater than other fuels. Despite these drawbacks, the high efficiency of liquid oxygen/liquid hydrogen makes these problems worth coping with when reaction time and storability are not too critical. Liquid hydrogen delivers a specific impulse about 30%-40% higher than most other rocket fuels.

Liquid oxygen and liquid hydrogen are used as the propellant in the high efficiency main engines of the Space Shuttle. LOX/LH2 also powered the upper stages of the Saturn V and Saturn 1B rockets, as well as the Centaur upper stage, the United States' first LOX/LH2 rocket (1962).

Another cryogenic fuel with desirable properties for space propulsion systems is liquid methane (-162 oC). When burned with liquid oxygen, methane is higher performing than state-of-the-art storable propellants but without the volume increase common with LOX/LH2 systems, which results in an overall lower vehicle mass as compared to common hypergolic propellants. LOX/methane is also clean burning and non-toxic. Future missions to Mars will likely use methane fuel because it can be manufactured partly from Martian in-situ resources. LOX/methane has no flight history and very limited ground-test history.

Liquid fluorine (-188 oC) burning engines have also been developed and fired successfully. Fluorine is not only extremely toxic; it is a super-oxidizer that reacts, usually violently, with almost everything except nitrogen, the lighter noble gases, and substances that have already been fluorinated. Despite these drawbacks, fluorine produces very impressive engine performance. It can also be mixed with liquid oxygen to improve the performance of LOX-burning engines; the resulting mixture is called FLOX. Because of fluorine's high toxicity, it has been largely abandoned by most space-faring nations.

Some fluorine containing compounds, such as chlorine pentafluoride, have also been considered for use as an 'oxidizer' in deep-space applications.

Hypergolic propellants are fuels and oxidizers that ignite spontaneously on contact with each other and require no ignition source. The easy start and restart capability of hypergols make them ideal for spacecraft maneuvering systems. Also, since hypergols remain liquid at normal temperatures, they do not pose the storage problems of cryogenic propellants. Hypergols are highly toxic and must be handled with extreme care.

Hypergolic fuels commonly include hydrazine, monomethyl hydrazine (MMH) and unsymmetrical dimethyl hydrazine (UDMH). Hydrazine gives the best performance as a rocket fuel, but it has a high freezing point and is too unstable for use as a coolant. MMH is more stable and gives the best performance when freezing point is an issue, such as spacecraft propulsion applications. UDMH has the lowest freezing point and has enough thermal stability to be used in large regeneratively cooled engines. Consequently, UDMH is often used in launch vehicle applications even though it is the least efficient of the hydrazine derivatives. Also commonly used are blended fuels, such as Aerozine 50 (or "50-50"), which is a mixture of 50%

UDMH and 50% hydrazine. Aerozine 50 is almost as stable as UDMH and provides better performance.

The oxidizer is usually nitrogen tetroxide (NTO) or nitric acid. In the United States, the nitric acid formulation most commonly used is type III-A, called inhibited red-fuming nitric acid (IRFNA), which consists of HNO3 + 14% N2O4 + 1.5-2.5% H2O + 0.6% HF (added as a corrosion inhibitor). Nitrogen tetroxide is less corrosive than nitric acid and provides better performance, but it has a higher freezing point. Consequently, nitrogen tetroxide is usually the oxidizer of choice when freezing point is not an issue, however, the freezing point can be lowered with the introduction nitric oxide. The resulting oxidizer is called mixed oxides of nitrogen (MON). The number included in the description, e.g. MON-3 or MON-25, indicates the percentage of nitric oxide by weight. While pure nitrogen tetroxide has a freezing point of about -9 oC, the freezing point of MON-3 is -15 oC and that of MON-25 is -55 oC.

USA military specifications for IRFNA were first published in 1954, followed in 1955 with UDMH specifications.

The Titan family of launch vehicles and the second stage of the Delta II rocket use NTO/Aerozine 50 propellant. NTO/MMH is used in the orbital maneuvering system (OMS) and reaction control system (RCS) of the Space Shuttle orbiter. IRFNA/UDMH is often used in tactical missiles such as the US Army's Lance (1972-91).

Hydrazine is also frequently used as a monopropellant in catalytic decomposition engines. In these engines, a liquid fuel decomposes into hot gas in the presence of a catalyst. The decomposition of hydrazine produces temperatures up to about 1,100 oC (2,000 oF) and a specific impulse of about 230 or 240 seconds. Hydrazine decomposes to either hydrogen and nitrogen, or ammonia and nitrogen.

Other propellants have also been used, a few of which deserve mentioning

Alcohols were commonly used as fuels during the early years of rocketry. The German V-2 missile, as well as the USA Redstone, burned LOX and ethyl alcohol (ethanol), diluted with water to reduce combustion chamber temperature. However, as more efficient fuels where developed, alcohols fell into general disuse.

Hydrogen peroxide once attracted considerable attention as an oxidizer and was used in Britain's Black Arrow rocket. In high concentrations, hydrogen peroxide is called high-test peroxide (HTP). The performance and density of HTP is close to that of nitric acid, and it is far less toxic and corrosive; however it has a poor freezing point and is unstable. Although HTP never made it as an oxidizer in large bi-propellant applications, it has found widespread use as a monopropellant. In the presence of a catalyst, HTP decomposes into oxygen and superheated steam and produces a specific impulse of about 150 s.

Nitrous oxide has been used as both an oxidizer and as a monopropellant. It is the oxidizer of choice for many hybrid rocket designs and has been used frequently in amateur high-powered rocketry. In the presence of a catalyst, nitrous oxide will decompose exothermically into nitrogen and oxygen and produce a specific impulse of about 170 s.

SOLID PROPELLANTS

Solid propellant motors are the simplest of all rocket designs. They consist of a casing, usually steel, filled with a mixture of solid compounds (fuel and oxidizer) that burn at a rapid rate, expelling hot gases from a nozzle to produce thrust. When ignited, a solid propellant burns from the center out towards the sides of the casing. The shape of the center channel determines the rate and pattern of the burn, thus providing a means to control thrust. Unlike liquid propellant engines, solid propellant motors cannot be shut down. Once ignited, they will burn until all the propellant is exhausted.

There are two families of solids propellants: homogeneous and composite. Both types are dense, stable at ordinary temperatures, and easily storable.

Homogeneous propellants are either simple base or double base. A simple base propellant consists of a single compound, usually nitrocellulose, which has both an oxidation capacity and a reduction capacity. Double base propellants usually consist of nitrocellulose and nitroglycerine, to which a plasticiser is added. Homogeneous propellants do not usually have specific impulses greater than about 210 seconds under normal conditions. Their main asset is that they do not produce traceable fumes and are, therefore, commonly used in tactical weapons. They are also often used to perform subsidiary functions such as jettisoning spent parts or separating one stage from another.

Modern composite propellants are heterogeneous powders (mixtures) that use a crystallized or finely ground mineral salt as an oxidizer, often ammonium perchlorate, which constitutes between 60% and 90% of the mass of the propellant. The fuel itself is generally aluminum. The propellant is held together by a polymeric binder, usually polyurethane or polybutadienes, which is also consumed as fuel. Additional compounds are sometimes included, such as a catalyst to help increase the burning rate, or other agents to make the powder easier to manufacture. The final product is rubber like substance with the consistency of a hard rubber eraser.

Composite propellants are often identified by the type of polymeric binder used. The two most common binders are polybutadiene acrylic acid acrylonitrile (PBAN) and hydroxy-terminator polybutadiene (HTPB). PBAN formulations give a slightly higher specific impulse, density, and burn rate than equivalent formulations using HTPB. However, PBAN propellant is the more difficult to mix and process and requires an elevated curing temperature. HTPB binder is stronger and more flexible than PBAN binder. Both PBAN and HTPB formulations result in propellants that deliver excellent performance, have good mechanical properties, and offer potentially long burn times.

Solid propellant motors have a variety of uses. Small solids often power the final stage of a launch vehicle, or attach to payloads to boost them to higher orbits. Medium solids such as the Payload Assist Module (PAM) and the Inertial Upper Stage (IUS) provide the added boost to place satellites into geosynchronous orbit or on planetary trajectories.

The Titan, Delta, and Space Shuttle launch vehicles use strap-on solid propellant rockets to provide added thrust at liftoff. The Space Shuttle uses the largest solid rocket motors ever built and flown. Each booster contains 500,000 kg (1,100,000 pounds) of propellant and can produce up to 14,680,000 Newtons (3,300,000 pounds) of thrust.

Hybrid Propellants

Hybrid propellant engines represent an intermediate group between solid and liquid propellant engines. One of the substances is solid, usually the fuel, while the other, usually the oxidizer, is liquid. The liquid is injected into the solid, whose fuel reservoir also serves as the combustion chamber. The main advantage of such engines is that they have high performance, similar to that of solid propellants, but the combustion can be moderated, stopped, or even restarted. It is difficult to make use of this concept for vary large thrusts, and thus, hybrid propellant engines are rarely built.

Abrasive

An abrasive is a material, often a mineral, that is used to shape or finish a workpiece through rubbing which leads to part of the workpiece being worn away. While finishing a material often means polishing it to gain a smooth, reflective surface it can also involve roughening as in satin, matte or beaded finishes.

Abrasives are extremely commonplace and are used very extensively in a wide variety of industrial, domestic, and technological applications. This gives rise to a large variation in the physical and chemical composition of abrasives as well as the shape of the abrasive. Common uses for abrasives include grinding, polishing, buffing, honing, cutting, drilling, sharpening, lapping, and sanding (see abrasive machining). (For simplicity, "mineral" in this article will be used loosely to refer to both minerals and mineral-like substances whether man-made or not.)

Files act by abrasion but are not classed as abrasives as they are a shaped bar of metal. However, diamond files are a form of coated abrasive (as they are metal rods coated with diamond powder).

Abrasives generally rely upon a difference in hardness between the abrasive and the material being worked upon, the abrasive being the harder of the two substances. However, this is not necessary as any two solid materials that repeatedly rub against each other will tend to wear each other away (such as softer shoe soles wearing away wooden or stone steps over decades or centuries or glaciers abrading stone valleys).

Typically, materials used as abrasives are either hard minerals (rated at 7 or above on Mohs scale of mineral hardness) or are synthetic stones, some of which may be chemically and physically identical to naturally occurring minerals but which cannot be called minerals as they did not arise naturally. (While useful for comparative purposes, the Mohs scale is of limited value to materials engineers as it is an arbitrary, ordinal, irregular scale.) Diamond, a common abrasive, for instance occurs both naturally and is industrially produced, as is corundum which occurs naturally but which is nowadays more commonly manufactured from bauxite. However, even softer minerals like calcium carbonate are used as abrasives, such as "polishing agents" in toothpaste.

These minerals are either crushed or are already of a sufficiently small size (anywhere from macroscopic grains as large as about 2 mm to microscopic grains about 0.001 mm in diameter) to permit their use as an abrasive. These grains, commonly called grit, have rough edges, often terminating in points which will decrease the surface area in contact and increase the localised contact pressure. The abrasive and the material to be worked are brought into contact while in relative motion to each other. Force applied through the grains causes fragments of the worked material to break away while simultaneously smoothing the abrasive grain and/or causing the grain to work loose from the rest of the abrasive.

Some factors which will affect how quickly a substance is abraded include:

- Difference in hardness between the two substances: a much harder abrasive will cut faster and deeper

- Grain size (grit size): larger grains will cut faster as they also cut deeper
- Adhesion between grains, between grains and backing, between grains and matrix: determines how quickly grains are lost from the abrasive and how soon fresh grains, if present, are exposed
- Contact force: more force will cause faster abrasion
- Loading: worn abrasive and cast off work material tends to fill spaces between abrasive grains so reducing cutting efficiency while increasing friction
- Use of lubricant/coolant/metalworking fluid: Can carry away swarf (preventing loading), transport heat (which may affect the physical properties of the workpiece or the abrasive), decrease friction (with the substrate or matrix), suspend worn work material and abrasives allowing for a finer finish, conduct stress to the workpiece.

Abrasive minerals

Abrasives may be classified as either natural or synthetic. When discussing sharpening stones, natural stones have long been considered superior but advances in material technology are seeing this distinction become less distinct. Many synthetic abrasives are effectively identical to a natural mineral, differing only in that the synthetic mineral has been manufactured rather than been mined. Impurities in the natural mineral may make it less effective.

Some naturally occurring abrasives are:

- Calcite (calcium carbonate)
- Emery (impure corundum)
- Diamond dust (synthetic diamonds are used extensively)
- Novaculite
- Pumice dust
- Rouge
- Sand

Some abrasive minerals (such as zirconia alumina) occur naturally but are sufficiently rare or sufficiently more difficult/costly to obtain such that a synthetic stone is used industrially. These and other artificial abrasives include:

- Borazon (cubic boron nitride or CBN)
- Ceramic

- Ceramic aluminium oxide
- Ceramic iron oxide
- Corundum (alumina or aluminium oxide)
- Dry ice
- Glass powder
- Steel abrasive
- Silicon carbide (carborundum)
- Zirconia alumina

Adhesive

An adhesive, also known as glue, is a material, typically liquid or semi-liquid, that adheres or bonds items together. Adhesives come from either natural or synthetic sources. The types of materials that can be bonded are vast but adhesives are especially useful for bonding thin materials. Adhesives cure (harden) by either evaporating a solvent or by chemical reactions that occur between two or more constituents.

Adhesives are useful for joining thin or dissimilar materials, minimizing weight, and providing a vibration-damping joint. A disadvantage of most adhesives is that most do not form an instantaneous joint, unlike many other joining processes, because the adhesive needs time to cure.

The earliest known date for a simple glue is 200,000 BC and for a compound glue 70,000 BC.

Adhesives are typically organized by the method of adhesion. These are then organized into reactive and non-reactive adhesives, which refers to if the adhesive chemically reacts to harden. Alternatively they can be organized by whether the raw stock is of natural, or synthetic origin, or by their starting physical phase.

Non-reactive adhesives

Drying adhesives

There are two types of adhesives that harden by drying: solvent based adhesives and polymer dispersion adhesives, also known as emulsion adhesives. Solvent based adhesives are a mixture of ingredients (typically polymers) dissolved in a solvent. White glue, contact adhesives and rubber cements are members of the drying adhesive family. As the solvent evaporates, the adhesive hardens. Depending on the chemical composition of the adhesive, they will adhere to different materials to greater or lesser degrees. Polymer dispersion adhesives are milky-white dispersions often based on polyvinyl acetate (PVAc). They are used extensively in the woodworking and packaging industries; also used with fabrics and fabric-based components, and in engineered products such as loudspeaker cones.

PRESSURE SENSITIVE ADHESIVES

Pressure sensitive adhesives (PSA) form a bond by the application of light pressure to marry the adhesive with the adherend. They are designed with a balance between flow and resistance to flow. The bond forms because the adhesive is soft enough to flow (i.e. "wet") to the adherend. The bond has strength because the adhesive is hard enough to resist flow when stress is applied to the bond. Once the adhesive and the adherend are in close proximity, molecular interactions, such as van der Waals forces, become involved in the bond, contributing significantly to its ultimate strength.

PSAs are designed for either permanent or removable applications. Examples of permanent applications include safety labels for power equipment, foil tape for HVAC duct work, automotive interior trim assembly, and sound/vibration damping films. Some high performance permanent PSAs exhibit high adhesion values and can support kilograms of weight per square centimeter of contact area, even at elevated temperature. Permanent PSAs may be initially removable (for example to recover mislabeled goods) and build adhesion to a permanent bond after several hours or days.

Removable adhesives are designed to form a temporary bond, and ideally can be removed after months or years without leaving residue on the adherend. Removable adhesives are used in applications such as surface protection films, masking tapes, bookmark and note papers, price marking labels, promotional graphics materials, and for skin contact (wound care dressings, EKG electrodes, athletic tape, analgesic and transdermal drug patches, etc.). Some removable adhesives are designed to repeatedly stick and unstick. They have low adhesion and generally can not support much weight.

Pressure sensitive adhesives are manufactured with either a liquid carrier or in 100% solid form. Articles are made from liquid PSAs by coating the adhesive and drying off the solvent or water carrier. They may be further heated to initiate a cross-linking reaction and increase molecular weight. 100% solid PSAs may be low viscosity polymers that are coated and then reacted with radiation to increase molecular weight and form the adhesive; or they may be high viscosity materials that are heated to reduce viscosity enough to allow coating, and then cooled to their final form. Major raw material for PSA's are acrylate based polymers.

Contact Adhesives

Contact adhesives are used in strong bonds with high shear-resistance like laminates, such as bonding Formica to a wooden counter, and in footwear, as in attaching outsoles to uppers.

Natural rubber and polychloroprene (Neoprene) are commonly used contact adhesives. Both of these elastomers undergo strain crystallization. In the construction industry a specialised proprietary adhesive known as Liquid Nails (or liquid nails as the generic), is used. This also copes with tasks such as sealing artificial turf.

Contact adhesives must be applied to both surfaces and allowed some time to dry before the two surfaces are pushed together. Some contact adhesives require as long as 24 hours to dry before the surfaces are to be held together. Once the surfaces are pushed together, the bond forms very quickly. It is usually not necessary to apply pressure for a long time, so there is less need for clamps.

Hot Adhesives

Hot adhesives, also known as hot melt adhesives, are thermoplastics applied in molten form (in the 65-180 °C range) which solidify on cooling to form strong bonds between a wide range of materials. Ethylene-vinyl acetate based hot-melts are particularly popular for crafts because of their ease of use and the wide range of common materials they can join. A glue gun (shown at right) is one method of applying hot adhesives. The glue gun melts the solid adhesive then allows the liquid to pass through its barrel onto the material, where it solidifies.

Thermoplastic glue may have been invented around 1940 by Procter & Gamble as a solution to water-based adhesives commonly used in packaging at that time failing in humid climates, causing packages to open.

Reactive adhesives

Multi-part adhesives

Multi-component adhesives harden by mixing two or more components which chemically react. This reaction causes polymers to cross-link into acrylics, urethanes, and epoxies.

There are several commercial combinations of multi-component adhesives in use in industry. Some of these combinations are:

- Polyester resin – polyurethane resin
- Polyols – polyurethane resin

- Acrylic polymers – polyurethane resins

The individual components of a multi-component adhesive are not adhesive by nature. The individual components react with each other after being mixed and show full adhesion only on curing. The multi-component resins can be either solvent-based or solvent-less. The solvents present in the adhesives are a medium for the polyester or the polyurethane resin. The solvent is dried during the curing process.

One-part adhesives

One-part adhesives harden via a chemical reaction with an external energy source, such as radiation, heat, and moisture.

Ultraviolet (UV) light curing adhesives, also known as light curing materials (LCM), have become popular within the manufacturing sector due to their rapid curing time and strong bond strength. Light curing adhesives can cure in as little as a second and many formulations can bond dissimilar substrates (materials) and withstand harsh temperatures. These qualities make UV curing adhesives essential to the manufacturing of items in many industrial markets such as electronics, telecommunications, medical, aerospace, glass, and optical. Unlike traditional adhesives, UV light curing adhesives not only bond materials together but they can also be used to seal and coat products. They are generally acrylic based.

Heat curing adhesives consist of a pre-made mixture of two or more components. When heat is applied the components react and cross-link. This type of adhesive includes epoxies, urethanes, and polyimides.

Moisture curing adhesives cure when they react with moisture present on the substrate surface or in the air. This type of adhesive includes cyanoacrylates and urethanes.

Natural adhesives

Natural adhesives are made from organic sources such as vegetable matter, starch (dextrin), natural resins or from animals e.g. casein or animal glue. They are often referred to as bioadhesives. One example is a simple paste made by cooking flour in water. Animal glues are traditionally used in bookbinding, wood joining, and many other areas but now are largely replaced by synthetic glues. Casein is mainly used to adhere glass bottle labels. Starch based adhesives are used in corrugated board production and paper sack production, paper tube winding, and wall paper adhesives. Masonite, a wood hardboard, was bonded using natural lignin, (although most modern MDF particle boards use synthetic thermosetting resins). Another form of natural adhesive is blood albumen (made from protein component of blood), which is used in the plywood industry. Animal glue remains the preferred glue of the luthier. Casein based glues are made by precipitating casein from milk protein using the acetic acid from vinegar. This forms curds, which are neutralized with a base, such as sodium bicarbonate (baking soda), to cause them to unclump and become a thicker plastic-like substance.

Explosive Material

An explosive material, also called an explosive, is a reactive substance that contains a great amount of potential energy that can produce an explosion if released suddenly, usually accompanied by the production of light, heat, sound, and pressure. An explosive charge is a measured quantity of explosive material.

This potential energy stored in an explosive material may be

- Chemical energy, such as nitroglycerin or grain dust
- Pressurized gas, such as a gas cylinder or aerosol can.
- Nuclear energy, such as in the fissile isotopes uranium-235 and plutonium-239

Explosive materials may be categorized by the speed at which they expand. Materials that detonate (explode faster than the speed of sound) are said to be "high explosives" and materials that deflagrate are said to be "low explosives". Explosives may also be categorized by their sensitivity. Sensitive materials that can be initiated by a relatively small amount of heat or pressure are primary explosives and materials that are relatively insensitive are secondary or tertiary explosives.

A wide variety of chemicals can explode; a smaller number are manufactured in quantity as explosives. The remainder are too dangerous, sensitive, toxic, expensive, unstable, or decompose too quickly for common usage.

Properties of explosive materials

To determine the suitability of an explosive substance for a particular use, its physical properties must first be known. The usefulness of an explosive can only be appreciated when the properties and the factors affecting them are fully understood. Some of the more important characteristics are listed below:

Availability and cost

The availability and cost of explosives are determined by the availability of the raw materials and the cost, complexity, and safety of the manufacturing operations.

Sensitivity

Sensitivity refers to the ease with which an explosive can be ignited or detonated, i.e., the amount and intensity of shock, friction, or heat that is required. When the term sensitivity is used, care must be taken to clarify what kind of sensitivity is under discussion. The relative sensitivity of a given explosive to impact may vary greatly from its sensitivity to friction or heat. Some of the test methods used to determine sensitivity relate to:

- Impact — Sensitivity is expressed in terms of the distance through which a standard weight must be dropped onto the material to cause it to explode.
- Friction — Sensitivity is expressed in terms of what occurs when a weighted pendulum scrapes across the material (it may snap, crackle, ignite, and/or explode).
- Heat — Sensitivity is expressed in terms of the temperature at which flashing or explosion of the material occurs.

Sensitivity is an important consideration in selecting an explosive for a particular purpose. The explosive in an armor-piercing projectile must be relatively insensitive, or the shock of impact would cause it to detonate before it penetrated to the point desired. The explosive lenses around nuclear charges are also designed to be highly insensitive, to minimize the risk of accidental detonation.

SENSITIVITY TO INITIATION

The index of the capacity of an explosive to be initiated into detonation in a sustained manner. It is defined by the power of the detonator which is certain to prime the explosive to a sustained and continuous detonation. Reference is made to the Sellier-Bellot scale that consists of a series of 10 detonators, from n. 1 to n. 10, each of which corresponds to an increasing charge weight. In practice, most of the explosives on the market today are sensitive to an n. 8 detonator, where the charge corresponds to 2 grams of mercury fulminate.

Velocity of Detonation

The velocity with which the reaction process propagates in the mass of the explosive. Most commercial mining explosives have detonation velocities ranging from 1800 m/s to 8000 m/s. Today, velocity of detonation can be measured with accuracy. Together with density it is an important element influencing the yield of the energy transmitted for both atmospheric overpressure and ground acceleration.

Stability

Stability is the ability of an explosive to be stored without deterioration.

The following factors affect the stability of an explosive:

- Chemical constitution. In the strictest technical sense, the word "stability" is a thermodynamic term referring to the energy of a substance relative to a reference state or to some other substance. However, in the context of explosives, stability commonly refers to ease of detonation, which is concerned with kinetics (i.e., rate of decomposition). It is perhaps best, then, to differentiate between the terms thermodynamically stable and kinetically stable by referring to the latter as "inert." Contrarily, a kinetically unstable substance is said to be "labile." It is generally recognized that certain groups like nitro (–NO2), nitrate (–ONO2), and azide (–N3), are intrinsically labile. Kinetically, there exists a low activation barrier to the decomposition reaction. Consequently, these compounds exhibit high sensitivity to flame or mechanical shock. The chemical bonding in these compounds is characterized as predominantly covalent and thus they are not thermodynamically stabilized by a high ionic-lattice energy. Furthermore, they generally have positive enthalpies of formation and there is little mechanistic hindrance to internal molecular rearrangement to yield the more thermodynamically stable (more strongly bonded) decomposition products. For example, in lead azide, Pb(N3)2, the nitrogen atoms are already bonded to one another, so decomposition into Pb and N2. is relatively easy.
- Temperature of storage. The rate of decomposition of explosives increases at higher temperatures. All standard military explosives may be considered to have a high degree of stability at temperatures from –10 to +35 °C, but each has a high temperature at which its rate of decomposition rapidly accelerates and stability is reduced. As a rule of thumb, most explosives become dangerously unstable at temperatures above 70 °C.
- Exposure to sunlight. When exposed to the ultraviolet rays of sunlight, many explosive compounds containing nitrogen groups rapidly decompose, affecting their stability.
- Electrical discharge. Electrostatic or spark sensitivity to initiation is common in a number of explosives. Static or other electrical discharge may be sufficient to cause a reaction, even detonation, under some circumstances. As a result, safe handling of explosives and pyrotechnics usually requires proper electrical grounding of the operator.

Power, performance, and strength

The term power or performance as applied to an explosive refers to its ability to do work. In practice it is defined as the explosive's ability to accomplish what is intended in the way of energy delivery (i.e., fragment projection, air blast, high-velocity jet, underwater shock and bubble energy, etc.). Explosive power or performance is evaluated by a tailored series of tests to assess the material for its intended use. Of the tests listed below, cylinder expansion and air-blast tests are common to most testing programs, and the others support specific applications.

- Cylinder expansion test. A standard amount of explosive is loaded into a long hollow cylinder, usually of copper, and detonated at one end. Data is collected concerning the rate of radial expansion of the cylinder and the maximum cylinder wall velocity. This also establishes the Gurney energy or 2E.
- Cylinder fragmentation. A standard steel cylinder is loaded with explosive and detonated in a sawdust pit. The fragments are collected and the size distribution analyzed.
- Detonation pressure (Chapman-Jouguet condition). Detonation pressure data derived from measurements of shock waves transmitted into water by the detonation of cylindrical explosive charges of a standard size.
- Determination of critical diameter. This test establishes the minimum physical size a charge of a specific explosive must be to sustain its own detonation wave. The procedure involves the detonation of a series of charges of different diameters until difficulty in detonation wave propagation is observed.
- Infinite-diameter detonation velocity. Detonation velocity is dependent on loading density (c), charge diameter, and grain size. The hydrodynamic theory of detonation used in predicting explosive phenomena does not include the diameter of the charge, and therefore a detonation velocity, for an imaginary charge of infinite diameter. This procedure requires the firing of a series of charges of the same density and physical structure, but different diameters, and the extrapolation of the resulting detonation velocities to predict the detonation velocity of a charge of infinite diameter.
- Pressure versus scaled distance. A charge of a specific size is detonated and its pressure effects measured at a standard distance. The values obtained are compared with those for TNT.

- Impulse versus scaled distance. A charge of a specific size is detonated and its impulse (the area under the pressure-time curve) measured as a function of distance. The results are tabulated and expressed as TNT equivalents.
- Relative bubble energy (RBE). A 5 to 50 kg charge is detonated in water and piezoelectric gauges measure peak pressure, time constant, impulse, and energy.

The RBE may be defined as Kx 3

RBE = Ks

where K = the bubble expansion period for an experimental (x) or a standard (s) charge.

Brisance

In addition to strength, explosives display a second characteristic, which is their shattering effect or brisance (from the French meaning to "break"), which is distinguished and separate from their total work capacity. This characteristic is of practical importance in determining the effectiveness of an explosion in fragmenting shells, bomb casings, grenades, and the like. The rapidity with which an explosive reaches its peak pressure (power) is a measure of its brisance. Brisance values are primarily employed in France and Russia.

The sand crush test is commonly employed to determine the relative brisance in comparison to TNT. No test is capable of directly comparing the explosive properties of two or more compounds; it is important to examine the data from several such tests (sand crush, trauzl, and so forth) in order to gauge relative brisance. True values for comparison require field experiments.

Density

Density of loading refers to the mass of an explosive per unit volume. Several methods of loading are available, including pellet loading, cast loading, and press loading, the choice being determined by the characteristics of the explosive. Dependent upon the method employed, an average density of the loaded charge can be obtained that is within 80–99% of the theoretical maximum density of the explosive. High load density can reduce sensitivity by making the mass more resistant to internal friction. However, if density is increased to the extent that individual crystals are crushed, the explosive may become more sensitive. Increased load density also permits the use of more explosive, thereby increasing the power of the warhead. It is possible to compress an explosive beyond a point of sensitivity, known also as dead-pressing, in which the material is no longer capable of being reliably initiated, if at all.

Volatility

Volatility is the readiness with which a substance vaporizes. Excessive volatility often results in the development of pressure within rounds of ammunition and separation of mixtures into their constituents. Volatility affects the chemical composition of the explosive such that a marked reduction in stability may occur, which results in an increase in the danger of handling.

Hygroscopicity and water resistance

The introduction of water into an explosive is highly undesirable since it reduces the sensitivity, strength, and velocity of detonation of the explosive. Hygroscopicity is used as a measure of a material's moisture-absorbing tendencies. Moisture affects explosives adversely by acting as an inert material that absorbs heat when vaporized, and by acting as a solvent medium that can cause undesired chemical reactions. Sensitivity, strength, and velocity of detonation are reduced by inert materials that reduce the continuity of the explosive mass. When the moisture content evaporates during detonation, cooling occurs, which reduces the temperature of reaction. Stability is also affected by the presence of moisture since moisture promotes decomposition of the explosive and, in addition, causes corrosion of the explosive's metal container.

Explosives considerably differ from one another as to their behavior in the presence of water. Gelatin dynamites containing nitroglycerine have a degree of water resistance. Explosives based on ammonium nitrate have little or no water resistance due to the reaction between ammonium nitrate and water, which liberates ammonia, nitrogen dioxide and hydrogen peroxide. In addition, ammonium nitrate is hygroscopic, susceptible to damp, hence the above concerns.

Toxicity

There are many types of explosives which are toxic to some extent. Manufacturing inputs can also be organic compounds or hazardous materials that require special handing due to risks (such as carcinogens). The decomposition products, residual solids or gases of some explosives can be toxic, whereas others are harmless, such as carbon dioxide and water. Examples of harmful by-products are:

- Heavy metals, such as lead, mercury and barium from primers (observed in high volume firing ranges).
- Nitric oxides from TNT.
- Perchlorates when used in large quantities.

Green explosives seek to reduce environment and health impacts. An example of such is the lead-free primary explosive Copper(I) 5-nitrotetrazolate, an alternative to lead azide.

General Properties of Engineering Materials:

The economists mainly concerned with the industries manufacturing engineering materials are interested in finding out ways in which the engineering materials can be used or applied in the best economic way. The economist should have working knowledge of properties and manufacturing processes of various engineering materials.

The services of such an economist can be of extreme use in making choice of engineering materials under specific conditions or in deciding the line of manufacture of engineering materials from the available local raw materials.

The fields of application of a particular engineering material are governed by the characteristics and various properties of that engineering material.

Such properties may be classified into various categories as follows:

(1) Chemical Properties:

The chemical properties of the material suggest the tendency of the material to combine with other substances, its reactivity, solubility and effects like corrosion, chemical composition, acidity, alkalinity, etc. Corrosion is one of the serious problems faced by engineers in selecting engineering materials, caused due to chemical properties of metal.

In metals, valence electrons are loosely bound to their atoms and can be easily removed during chemical reactions. So, when metals are exposed in the atmosphere and come in contact with gases like oxygen, chlorine etc., the chemical reactions take place. When iron reacts with oxygen, iron oxide is formed which is red in colour and iron metal is coated with it. This is called corrosion.

(2) Electrical Properties:

These properties signify the ability of the material to resist the flow of an electric current and they include conductivity, dielectric strength and resistivity.

(3) Magnetic Properties:

The study of magnetic properties of the material like permeability, hysteresis and coercive force is required when.it is to be used for generators, transformers, etc.

(4) Mechanical Properties:

The characteristics governing the behaviour of material when external forces are applied are included in these properties. Some of the important mechanical properties are elasticity, hardness, plasticity, strength, etc.

(5) Optical Properties:

When the material is to be used for the optical work, the knowledge of its optical properties like colour, light transmission, refractive index, reflectivity, etc. is necessary. When light strikes any material, it interacts with its atoms and causes various types of effects. The light may be reflected, refracted, scattered or absorbed. The study of light in materials and how to use this behaviour to control the various light effects is called optics.

(6) Physical Properties:

These are required to evaluate the condition of the material without any external force acting on it and they include bulk density, durability, porosity, etc.

(7) Thermal Properties:

The knowledge of thermal properties of the material like specific heat, thermal expansion and conductivity is helpful in knowing the response of the material to the thermal changes. Thus suitable materials can be selected to withstand fluctuating and high temperatures.

(8) Technological Properties:

The properties of metals and alloys which have a bearing on their processing or application are called technological properties. Castability, machinability, weldability and workability are some of the significant technological properties of metals and alloys.

Out of all such properties, the physical properties and mechanical properties are particularly very important to a construction engineer.

Physical Properties of Engineering Materials

Following terms in connection with the physical properties of engineering materials are defined and explained:

(1) Bulk density

(2) Chemical resistance

(3) Coefficient of softening

(4) Density

(5) Density index

(6) Durability

(7) Fire resistance

(8) Frost resistance

(9) Hygroscopicity

(10) Porosity

(11) Refractoriness

(12) Spalling resistance

(13) Specific heat

(14) Thermal capacity

(15) Thermal conductivity

(16) Water absorption

(17) Water permeability

(18) Weathering resistance.

(1) Bulk Density:

The term bulk density is used to mean the mass of a unit volume of material in its natural state i.e. including pores and voids. It is obtained by finding out the ratio of mass of specimen to the volume of specimen in its natural state.

The technical properties of the material such as strength, heat, conductivity, etc. are greatly influenced by its bulk density and hence the performance efficiency of a material will depend upon its bulk density.

For most of the materials, the bulk density is less than its density except for dense materials, liquids and materials obtained from the molten masses.

Table 1-1 shows the bulk densities of some of the important building materials.

(2) Chemical Resistance:

The ability of material to withstand the action of acids, alkalies, gases and salt solutions is known as its chemical resistance.

This property is carefully examined while selecting material for sewer pipes, hydraulic engineering installations, sanitary facilities, etc.

(3) Coefficient of Softening:

The ratio of compressive strength of material saturated with water to that in dry state is known as the coefficient of softening. The materials such as glass and metal are not affected by the presence of water and their coefficient of softening is unity. On the other hand, the materials like clay easily loose their strength when soaked in water and hence, their coefficient of softening is zero.

The materials having coefficient of softening equal to 0.8 or more are referred to as the water-resisting materials. It is advisable to avoid the use of materials with coefficient of softening less than 0.8 for situations which are likely to be exposed permanently to the action of moisture.

(4) Density:

The term density of a material is defined as the mass of a unit volume of homogeneous material. It is obtained by working out the ratio of mass of material to the volume of material in homogeneous state. The physical properties of a material are greatly influenced by its density.

(5) Density Index:

The ratio of bulk density of a material to its density is known as its density index and it thus denotes the degree to which its volume is filled up with solid matter.

As there are practically no dense substances in nature, the density index of most of the building materials is less than unity.

(6) Durability:

The property of a material to resist the combined action of atmospheric and other factors is known as its durability.

The running or maintenance cost of a building will naturally depend upon the durability of the materials of which it is composed.

(7) Fire Resistance:

The term fire resistance is used to mean the ability of a material to resist the action of high temperature without losing its load-bearing capacity i.e., without substantial loss of strength or deformation in shape.

This property of a material is of great importance in case of a fire and as the operation of fire-fighting is usually accompanied by water, this property of a material is tested by the combined actions of high temperature and water. The material should be sufficiently fireproof to afford safety and stability in case of a fire.

(8) Frost Resistance:

The ability of a water-saturated material to resist repeated freezing and thawing without considerable decrease of mechanical strength or visible signs of failure is known as the frost resistance. The frost resistance of a material depends upon the density of material and its degree of saturation with water.

In general, the dense materials are frost resistant. The porous materials whose pores are closed or filled with water to less than 90% of their volume are frost resistant.

(9) Hygroscopicity:

The property of a material to absorb water vapour from air is known as the hygroscopicity and it is governed by the nature of substance involved, number of pores, air temperature, relative humidity, etc. The water-retaining or hydrophilic substances readily dissolve in water.

(10) Porosity:

The term porosity is used to indicate the degree by which the volume of a material is occupied by pores. It is expressed as a ratio of volume of pores to that of the specimen. The porosity of a material is indicative of its various properties such as strength, bulk density, water absorption, thermal conductivity, durability, etc., and hence it is to be carefully studied and analysed.

(11) Refractoriness:

The ability of a material to withstand prolonged action of high temperature without melting or loosing shape is known as its refractoriness.

(12) Spalling Resistance:

The ability of a material to endure a certain number of cycles of sharp temperature variations without failing is known as its spalling resistance and it mainly depends on the coefficients of linear expansion of its constituents.

(13) Specific Heat:

The term specific heat is defined as the quantity of heat, expressed in kilocalories, required to heat 1 N of material by 1°C. The specific heat of a material is to be considered when heat accumulation is to be taken into account.

The specific heats of steel, stone and wood are as follows:

Steel – 0.046 x 103 J/N °C

Stone – 0.075 to 0.09 x 103 J/N °C

Wood – 0.239 to 0.27 x 103 J/N °C.

(14) Thermal Capacity:

The property of a material to absorb heat is known as its thermal capacity and it is worked out by the following equation –

T = H / M (T2 – T1)

Where, T = Thermal capacity in J/N °C

H = Quantity of heat required to increase the temperature of material from T1 to T2 in J M = Mass of material in N

T1 = Temperature of material before heating in °C

T2 = Temperature of material after heating in °C.

(15) Thermal Conductivity:

The thermal conductivity of a material is the amount of heat in kilocalories that will flow through unit area of the material with unit thickness in unit time when difference of temperature on its faces is also unity. The unit of thermal conductivity is J per m hr °C and it is usually denoted by K. The thermal conductivity of a material depends on its density, porosity, moisture content and temperature.

The term thermal resistivity of a material is used to mean the reciprocal of its thermal conductivity. The thermal resistance of a material is equal to thermal resistivity multiplied by its thickness.

(16) Water Absorption:

The ability of a material to absorb and retain water is known as its water absorption. The dry material is fully immersed in water and then the water absorption is worked out either as percentage of weight or percentage of volume of dry material. It mainly depends on the volume, size and shape of pores, present in the material.

(17) Water Permeability:

The capacity of a material to allow water to pass through it under pressure is known as its water permeability and it is described as the quantity of water that will pass through the material in one hour at constant pressure, the cross-sectional area of the specimen being 1 cm. The dense materials like glass, steel, etc. are water-proof or impervious to the water.

(18) Weathering Resistance:

The term weathering resistance is used to express the ability of a material to resist alternating wet and dry conditions without seriously affecting its shape and mechanical strength. It thus, indicates the behaviour of materials when exposed to changing conditions of humidity.

Mechanical Properties of Engineering Materials

The mechanical properties of materials like their rigidity, ductility and strength are of vital importance in determining their fabrication and possible practical applications.

The building materials show a wide range of the mechanical properties ranging from hardness of diamond to the ductility of pure copper and the astonishing elastic behaviour of rubber. In a similar way, many materials behave quite differently when stressed in different ways. For instance, cast-iron, cement and bricks are much stronger in compression whereas wood and steel are stronger in tension.

Following terms in connection with the common mechanical properties of building materials are defined and explained:

(1) Abrasion

(2) Creep

(3) Elasticity

(4) Fatigue

(5) Hardness

(6) Impact strength

(7) Plasticity and brittleness

(8) Strength

(9) Wear.

(1) Abrasion:

The resistance of a material to the abrasion is found out by dividing the difference in weights of specimens prior to and after abrasion with the area of abrasion.

(2) Creep:

In many applications, the building materials are required to sustain steady loads for long periods. Under such conditions, the material may continue to deform until its usefulness is seriously reduced. Such time-dependent deformations of a structure can grow large and may even result in final fracture without any increase in load. If the deformation continues even when the load is constant, such additional deformation is known as the creep.

Most of the building materials creep to a certain extent at all temperatures. However the engineering metals such as steel, aluminium and copper creep very little at room temperature. The high temperatures lead to rapid creep which is often accompanied by microstructural changes. The phenomenon of creep is important in polymers at room temperature, in alloys of aluminium at 100°C and in steels above 300°C.

(3) Elasticity:

When a load is applied to a material, there is change in its shape and dimension. The term elasticity is used to indicate the ability of a material to restore its initial form and dimensions after the load is removed.

The difference between the following two terms should be noted:

(i) Elastic Deformation: A deformation is said to be elastic when the solid deforms when it is loaded but returns to its original position when unloaded. A change in pressure or an application of load results in the elastic deformation. The term ideal deformation is used to mean the deformation that takes place instantaneously upon application of force and disappears completely on removal of the force. Such deformations obey Hooke's law and the elastic strain of the metal is directly proportional to the applied force. The ideal deformation occurs with comparatively smaller deformation forces which can keep the working stresses within the elastic limit.

(ii) Plastic Deformation: A deformation is said to be plastic when the solid retains full or partly the change in shape after the load is removed. The plastic deformation is observed when the stress exceeds the elastic limit and its rate is controlled by the strain rate, applied stress and temperature. It can occur under tensile, compressive and torsional stresses. It is intentionally carried out in processes like rolling, forging, etc. so as to make useful products.

(4) Fatigue:

When the materials are subjected to a repetitive or fluctuating stress, they will fail at a stress much lower than that required to cause fracture under steady loads.

This behaviour is called the fatigue and it is distinguished by the following three features:

(i) Increased uncertainty in strength and service life; (ii) Loss of ductility; and (iii) Loss of strength.

Following are the reasons of the fatigue failures:

(i) Corroding environments resulting in the reduction of the fatigue strength;

(ii) Stress concentration points;

(iii) Surface imperfections like machining marks and surface irregularities; and

(iv) Temperature, the fatigue strength being high at low temperatures and decreasing gradually with the rise in temperature.

(5) Hardness:

The ability of a material to resist penetration by a harder body is known as its hardness. It is a major factor in deciding the workability and use of a material for floors and road surfaces. The hardness is not a fundamental property. But it is a combined effect of compressive, elastic and plastic properties relative to the mode of penetration, shape of penetrator, etc.

The hardness bears a fairly constant relationship to the tensile strength of a given material. It can therefore be used as a practical non-destructive test to get roughly an idea of the tensile strength of the material and the state of the metal near the surface.

The hardness of stone materials can be determined with the help of Mohs' scale of hardness. It is a list of ten materials arranged in the order of increasing hardness. The hardness of a material lies between the hardness of two materials i.e. the one which scratches and the other which is scratched by the material to be tested.

(6) Impact Strength:

The impact strength of a material is the quantity of work required to cause its failure per its unit volume. It thus indicates the toughness of a material and the materials are tested in an impact testing machine to determine their impact strength.

The impact strength is a complex characteristic which takes into account both the toughness and strength of a material.

It varies with the following factors:

(i) If the dimensions of the specimen are increased, there is also increase in the impact strength.

(ii) If the sharpness of the notch increases, the impact strength required to cause failure decreases.

(iii) The angle of the notch also improves the impact strength after certain values.

(iv) The impact strength is also affected to a certain extent by the velocity of impact.

(v) The temperature of the specimen under test gives an indication about the type of fracture that is likely to occur i.e. ductile, brittle or ductile to brittle transition.

(7) Plasticity and Brittleness:

The term plasticity of a material is defined as its ability to change its shape under load without cracking and to retain its shape after the removal of load.

The materials can broadly be divided into two groups, namely, plastic materials and brittle materials. The steel, copper, hot bitumen, etc. are plastic materials. The brittle materials fail suddenly under pressure without appreciable deformation preceding the failure. The rock materials, ceramic materials, glass, cast-iron, concrete and some other materials are brittle and they offer poor resistance to bending, impact and tension.

(8) Strength:

The ability of a material to resist failure under the action of stresses caused by a load is known as its strength. The loads to which a material is commonly subjected to are compression, tension and bending. The corresponding strength is obtained by dividing the ultimate load with the cross-sectional area of the specimen.

The stresses in the building materials are not allowed to exceed a certain percentage of their ultimate strength. Thus, a margin of safety is provided and the term factor of safety is used to denote the ratio of ultimate stress to safe stress. For instance, if the factor of safety is two, the stress to be adopted for design purposes would be one-half of the ultimate stress.

The values of factors of safety are specified by design standards and they are framed by taking into account various factors such as nature of work, quality of material, service conditions, economic considerations, etc.

(9) Wear:

The failure of a material under the combined actions of abrasion and impact is known as its wear. The wear resistance is usually expressed as a percentage of loss in weight and it is of great importance in deciding the suitability of a material for use of road surfaces, railway ballast, etc.

2

HIGH POLYMERS FOR ENGINEERING APPLICATIONS

Several monomers combine chemically to form three-dimensional networks or a chain of elements, and this process is known as Polymerization. These monomers can be the same or different compounds combined to form a polymer. Besides, for a polymer to exhibit unique characteristics, it should be formed as a result of the combination of a minimum of 100 monomers. Usually, one molecule of polymer consists of hundreds and thousands of monomers.

A stable covalent bond is formed in the case of polymerization, and it helps the newly formed Polymers to have unique traits. This includes high tensile strength, elasticity, capability to form fibers.

Engineering polymers are a type of plastic materials that are durable and rugged. Because of these properties, this type of polymer is widely used in the manufacturing industry. It is also commonly used as material in consumer goods too.

Unlike commodity plastics, engineering plastics have the ability to handle heat and stress which makes them a perfect material for making containers and packaging. These polymers are also called thermoplastics. They are used by manufacturing companies as raw materials. They are subjected to high temperatures, melted and moulded to make usable products.

There are basically five types of engineering polymers. They are polyphenel, polyester, polyoxmethylene, polymide and polycarbonate. PolyVisions Inc. has made a mark in the science of polymers. With its nearly 30 years of operation, the company has worked with virtually all types of plastics including these types of engineering polymers.

HEAT GENERATING POLYMERS

Known for its high temperature threshold, Polyphenel is a common raw material in the plastic engineering industry. It is used to make heat generating consumer products including coffee makers, hair dryers and washing machines among many others. It is a tough material that can withstand heat repeatedly without melting. As such, polyphenel is a type of engineering plastics used for making ovenware and other food packaging and processing materials.

PolyVisions Inc. does not just work with polyphenel. It has found better ways to make food-safe products. The company has created what they call grafted PET resins which have top notch impact resistance. They are also capable of resisting and withstanding a wide range of temperatures. It is what makes PolyVisions a top choice in the food industry. Their products are ideal for freezing and industrial cooking applications.

Polyester Engineering Polymers

Polyester is not only useful in the garment industry. In fact, it can be just as useful to the electronics industry. It actually has a unique chemical makeup that has a high insulation rate. It is utilized in making ignition coils, automobile sections boards and televisions.

Supersteel Polymer

A plastic material with incredible strength, polyoxmethylene is dubbed as super steel. While it is strong engineering polymer, it is amazingly lightweight and sensitive too which makes it ideal for making mechanical instruments and gears.

Metal-Like Polymers

There are two types of engineering polymers that possess metal-like strength. These are polycarbonate and polymide. Polymide is highly useful in the automobile industry. It serves as a competent metal replacement in manufacturing fan blades, valves, dashboards and automobile bumpers.

While polymide is commonly used material in automobiles, polycarbonate is highly beneficial to the technology industry. This polymer has found its way to DVD players and computer parts. It is also utilized in making airplane cockpit glass and motorcycle helmets among others.

Polyvisions Inc. does not just work with these polymers. Their knowledge and experience in polymer engineering and science allows them to come up with effective solutions to even the most challenging polymer engineering problems.

The company's outstanding record in working with several industries including the healthcare, transportation, wire and cable industries among others is a testament to their mastery of engineering polymers.

Polymers may sound like a very industry specific thing, and to some degree they are, but they are actually a pretty big part of a lot of industries you may indeed recognize. Polymer testing and consultancy for plastic has applications in such industries as aerospace, automotive, electronics, packaging and medical devices. Polymers are incredibly diverse elements that represent such fields of engineering from avionics through biomedical applications, drug delivery system, biosensor devices, tissue engineering, cosmetics etc. the application of polymers and their subsequent composites is still advancing and increasing quickly due to their ease regarding manufacturing. When considering a polymer application, you will want to understand how the material behaves over time so you can assess its real value. It is important to understand that polymeric materials may include: raw materials, polymer compounds, foams, structural adhesives and composites, fillers, fibres, films, membranes, emulsions, coatings, rubbers, sealing materials, adhesive resins, solvents, inks and pigments. Keep in mind that these are some of the industries in which you would see the use and application of various polymeric materials and polymers themselves:

In aircraft, aerospace, and sports equipment

3D printing plastics

Biopolymers in molecular recognition

Polymers in holography

Organic polymers used in water purification

Printed circuit board substrates

Green Chemicals: Polymers and Biopolymers

Polymeric Biomolecules

Renewable Biomass operations

Polymers that are used in operation of bulletproof vests and fire-resistant jackets

Monomeric Units

To understand what these industries have in common, it may be good to go back to some basic knowledge. This includes- what a polymer is in the first place. Polymers are large molecules that are comprised of thousands or in some cases millions of atoms which are bonded in a pattern that repeats itself. You can imagine the structure of a polymer by simply imagining the image of a chain. The atoms are bonded together as the links in a chain are. In this way, they create a polymer chain. The molecular links in the polymer chain are often referred to as what are repeat units. They are formed by what are called monomers. Again, the structure of the repeat unit may vary widely. So what does it depend on? Well, it depends on the various raw materials that may comprise a polymer. Polyethylene is an example of a material that is used to create polymers as it is used to create a wide variety of plastic bags and containers.

No matter what industry your polymers are needed for, you can make sure you are getting the highest quality of (acrylic) polymers with Gellner Industrial. Gellner Industrial, LLC is the premier manufacturer of water based acrylic polymers, bringing over a quarter century of industry experience to our client base. Our vast product knowledge and specialized manufacturing processes ensure complete product conformance and quality assurance.

Polymer engineering is generally an engineering field that designs, analyses, and modifies polymer materials. Polymer engineering covers aspects of the petrochemical industry, polymerization, structure and characterization of polymers, properties of polymers, compounding and processing of polymers and description of major polymers, structure property relations and applications.

The word "polymer" was introduced by the Swedish chemist J. J. Berzelius. He considered, for example, benzene (C6H6) to be a polymer of ethyne (C2H2). Later, this definition underwent a subtle modification.

The history of human use of polymers has been long since the mid-19th century, when it entered the chemical modification of natural polymers. In 1839, C. Goodyear has found a critical advance in the research of rubber vulcanization, which has turned natural rubber into a practical engineering material. In 1870, J. W. Hyatt uses camphor to plasticize nitrocellulose to make nitrocellulose plastics industrial. 1907 L. Baekeland reported the synthesis of the first thermosetting phenolic resin, which was industrialized in the 1920s, the first synthetic plastic product. In 1920, H. Standinger proposed that polymers are long-chain molecules that are connected by structural units through common covalent bonds. This conclusion laid the foundation for the establishment of modern polymer science. Subsequently, Carothers divided the synthetic polymers into two broad categories, namely a polycondensate obtained by a polycondensation reaction and an addition polymer obtained by a polyaddition reaction. In the 1950s, K. Ziegler and G. Natta discovered a coordination polymerization catalyst and pioneered the era of synthesis of stereoregular polymers. In the decades after the establishment of the concept of macromolecules, the synthesis of high polymers has achieved rapid development, and many important polymers have been industrialized one after another.

Classification

The basic division of polymers into thermoplastics, elastomers and thermosets helps define their areas of application.

Thermoplastics

Thermoplastic refers to a plastic that has heat softening and cooling hardening properties. Most of the plastics we use in our daily lives fall into this category. It becomes soft and even flows when heated, and the cooling becomes hard. This process is reversible and can be repeated. Thermoplastics have relatively low tensile moduli, but also have lower densities and properties such as transparency which make them ideal for consumer products and medical products. They include polyethylene, polypropylene, nylon, acetal resin, polycarbonate and PET, all of which are widely used materials.

Elastomers

An elastomer generally refers to a material that can be restored to its original state after removal of an external force, whereas a material having elasticity is not necessarily an elastomer. The elastomer is only deformed under weak stress, and the stress can be quickly restored to a polymer material close to the original state and size. Elastomers are polymers which have very low moduli and show reversible extension when strained, a valuable property for vibration absorption and damping. They may either be thermoplastic (in which case they are known as Thermoplastic elastomers) or crosslinked, as in most conventional rubber products such as tyres. Typical rubbers used conventionally include natural rubber, nitrile rubber, polychloroprene, polybutadiene, styrene-butadiene and fluorinated rubbers.

Thermosets

A thermosetting resin is used as a main component, and a plastic which forms a product is formed by a cross-linking curing process in combination with various necessary additives. It is liquid in the early stage of the manufacturing or molding process, and it is insoluble and infusible after curing, and it cannot be melted or softened again. Common thermosetting plastics are phenolic plastics, epoxy plastics, aminoplasts, unsaturated polyesters, alkyd plastics, and the like. Thermoset plastics and thermoplastics together constitute the two major components of synthetic plastics. Thermosetting plastics are divided into two types: formaldehyde cross-linking type and other cross-linking type.

Thermosets includes phenolic resins, polyesters and epoxy resins, all of which are used widely in composite materials when reinforced with stiff fibers such as fiberglass and aramids. Since crosslinking stabilises the thermoset polymer matrix of these materials, they have physical properties more similar to traditional engineering materials like steel. However, their very much lower densities compared with metals makes them ideal for lightweight structures. In addition, they suffer less from fatigue, so are ideal for safety-critical parts which are stressed regularly in service.

Materials

Plastic

Plastic is a polymer compound which is polymerized by polyaddition polymerization and polycondensation. It is free to change the composition and shape. It is made up of synthetic resins and fillers, plasticizers, stabilizers, lubricants, colorants and other additives. The main component of plastic is resin. Resin means that the polymer compound has not been added with various additives. The term resin was originally named for the secretion of oil from plants and animals, such as rosin and shellac. Resin accounts for approximately 40% - 100% of the total weight of the plastic. The basic properties of plastics are mainly determined by the nature of the resin, but additives also play an important role. Some plastics are basically made of synthetic resins, with or without additives such as plexiglass, polystyrene, etc.

Fiber

Fiber refers to a continuous or discontinuous filament of one substance. Animals and plant fibers play an important role in maintaining tissue. Fibers are widely used and can be woven into good threads, thread ends and hemp ropes. They can also be woven into fibrous layers when making paper or feel. They are also commonly used to make other materials together with other materials to form composites. Therefore, whether it is natural or synthetic fiber filamentous material. In modern life, the application of fiber is ubiquitous, and there are many high-tech products.

Glass Fibers. Glass fibers are the most commonly used and cheapest of the engineering fibers. These materials are not new (they were first manufactured in 1893) but many of their structural applications are still developing. They are characterized by their strength, inertness, non-flammability and resistance to heat, chemicals and biological degradation. In their current form, they were first produced in the 1930's, and by 1950 developed into a reinforcement for plastics, currently their fastest growing application. E-glass fiber (a calciumalumina-borosilicate) is the predominant reinforcement for polymers because of its combination of good mechanical and electrical properties and low cost. Commercial formulations have a tensile strength of 300,000 to 450,000 psi (figure 3.4). Its specific tensile strength is double that of steel and over four times that of Aluminum.

Carbon Fibers. Carbon fibers are fine filaments composed largely of elemental carbon (up to 99.99+). These fibers are notable because of their high strength and stiffness (figure 3.4). Their stiffness can vary from 5 million psi (half of glass fiber) to 100 million psi (three times the stiffness of steel). They are also very light, chemically inert, with good electrical conductivity and low coefficient of thermal expansion. Carbon fibers are manufactured through the pyrolysis of carbon substrates such as rayon, polyacrylonitrile (PAN) or petroleum pitch. They are most commonly used as reinforcing elements of engineering polymers and other synthetic matrices. Most of the carbon fibers used in engineering applications are PAN based because these exhibit higher tensile strength while maintaining good stiffness. It is not yet possible to manufacture pitch based carbon fibers with mechanical properties comparable to PAN based fibers (pare 5.10). Carbon fibers are combined with thermoplastics and thermosets to form composite structures analogous to glass fiber composites but exhibiting higher strength and stiffness, at a higher cost.

Arachnid Fibera. Aramids, are a family of aromatic polyamide fibers. Aramids were first manufactured by DuPont in 1965. These fibers exhibit very high heat resistance and are therefore used in electrical and insulating applications. Aramid fibers are used in composites to provide lightweight and stiffness beyond what could be achieved through the use of glass. Aramid based composites are resistant to fatigue, impact damage and reduce vibration damping and wear. Because of their high cost (higher than glass or carbon fibers), aramid fibers have been adopted in systems where performance was the overriding design criteria (aero-space, defense applications); however, lately, these fibers have also been used as reinforcements for tires, conveyors and belts.

CERAMIC FIBERS

Ceramic fibers are non-metallic, inorganic filaments, of relative recent synthesis. The most common fibers are alumina, boron and silicon carbides used mostly because of their refractory properties. These fibers are also beginning to be used as reinforcements for plastics in composite structures. Their thermal stability, resistance to moisture, and corrosion resistance make them well suited to high temperature applications. Ceramic fibre reinforced polymers have been found to retain high performance mechanical properties after prolonged exposures to high temperature. However, the high cost of these composites prevent their use in bulk applications.

Ceramics are defined as inorganic non-metallic materials processed or consolidated at high temperatures. Advanced ceramics are materials characterized by a fine structure composed of numerous crystal particles and where the main constituents are Silicon, Aluminum or Zirconia. They are distinguished from more traditional ceramics by the specialized properties and sophisticated processing requirements. The manufacture of advanced ceramics involve chemical or physical treatment of sands, milling of the raw materials, forming either through pressing, extrusion or injection molding, sintering and bonding of the final product. Ceramics can be classified in several groups according to their chemical composition, oxides of alumina based ceramics are the most widely used with applications in the electronics and industrial machine sectors. Other oxides include beryllia and zirconia. In addition, Silicon Carbide and Nitrides are used in many thermal and wear resistance applications. Other commonly used ceramics include oxide alloys, ceramic composites with metals and ticania ceramics.

Rubber

Rubber refers to highly elastic polymer materials and reversible shapes. It is elastic at room temperature and can be deformed with a small external force. After removing the external force, it can return to the original state. Rubber is a completely amorphous polymer with a low glass transition temperature and a large molecular weight, often greater than several hundred thousand. Highly elastic polymer compounds can be classified into natural rubber and synthetic rubber. Natural rubber processing extracts gum rubber and grass rubber from plants; synthetic rubber is polymerized by various monomers. Rubber can be used as elastic, insulating, water-impermeable air-resistant materials.

Applications

Polyethylene

Commonly used polyethylenes can be classified into low density polyethylene (LDPE), high density polyethylene (HDPE), and linear low density polyethylene (LLDPE). Among them, HDPE has better thermal, electrical and mechanical properties, while LDPE and LLDPE have better flexibility, impact properties and film forming properties. LDPE and LLDPE are mainly used for plastic bags, plastic wraps, bottles, pipes and containers; HDPE is widely used in various fields such as film, pipelines and daily necessities because its resistance to many different solvents.

Polypropylene

Polypropylene is widely used in various applications due to its good chemical resistance and weldability. It has lowest density among commodity plastics. It is commonly used in packaging applications, consumer goods, automatic applications and medical applications. Polypropylene sheets are widely used in industrial sector to produce acid and chemical tanks, sheets, pipes, Returnable Transport Packaging (RTP), etc. because of its properties like high tensile strength, resistance to high temperatures and corrosion resistance.

Composites

Typical uses of composites are monocoque structures for aerospace and automobiles, as well as more mundane products like fishing rods and bicycles. The stealth bomber was the first all-composite aircraft, but many passenger aircraft like the Airbus and the Boeing 787 use an increasing proportion of composites in their fuselages, such as hydrophobic melamine foam. The quite different physical properties of composites gives designers much greater freedom in shaping parts, which is why composite products often look different from conventional products. On the other hand, some products such as drive shafts, helicopter rotor blades, and propellers look identical to metal precursors owing to the basic functional needs of such components.

Biomedical applications

Biodegradable polymers are widely used materials for many biomedical and pharmaceutical applications. These polymers are considered very promising for controlled drug delivery devices. Biodegradable polymers also offer great potential for wound management, orthopaedic devices, dental applications and tissue engineering. Not like non biodegradable polymers, they won't require a second step of a removal from body. Biodegradable polymers will break down and are absorbed by the body after they served their purpose. Since 1960, polymers prepared from glycolic acid and lactic acid have found a multitude of uses in the medical industry. Polylactates (PLAs) are popular for drug delivery system due to their fast and adjustable degradation rate.

Membrane technologies

Membrane techniques are successfully used in the separation in the liquid and gas systems for years, and the polymeric membranes are used most commonly because they have lower cost to produce and are easy to modify their surface, which make them suitable in different separation processes. Polymers helps in many fields including the application for separation of biological active compounds, proton exchange membranes for fuel cells and membrane contractors for carbon dioxide capture process.

Bulk polymerization

Bulk polymerization or mass polymerization is carried out by adding a soluble initiator to a pure monomer in liquid state. The initiator should dissolve in the monomer. The reaction is initiated by heating or exposure to radiation. As the reaction proceeds the mixture becomes more viscous. The reaction is exothermic and a wide range of molecular masses are produced.

Techniques of polymerization

Bulk polymerization is carried out in the absence of any solvent or dispersant and is thus the simplest in terms of formulation. It is used for most step-growth polymers and many types of chain-growth polymers. In the case of chain-growth reactions, which are generally exothermic, the heat evolved may cause the reaction to become too vigorous and difficult to control unless efficient cooling. Bulk polymerization has several advantages over other methods, these advantages are

The system is simple and requires thermal insulation.

The polymer obtained is pure.

Large castings may be prepared directly.

Molecular weight distribution can be easily changed with the use of a chain transfer agent.

The product obtained has high optical clarity

Disadvantages

Heat transfer and mixing become difficult as the viscosity of reaction mass increases.

The problem of heat transfer is compounded by the highly exothermic nature of free radical addition polymerization.

The polymerization is obtained with a broad molecular weight distribution due to the high viscosity and lack of good heat transfer.

Very low molecular weights are obtained.

Gel effect should be there

For reducing the disadvantages of bulk polymerization, the process can be carried out in a solution. This is known as solution polymerization.

Polymer characterization is the analytical branch of polymer science.

The discipline is concerned with the characterization of polymeric materials on a variety of levels. The characterization typically has as a goal to improve the performance of the material. As such, many characterization techniques should ideally be linked to the desirable properties of the material such as strength, impermeability, thermal stability, and optical properties.

Characterization techniques are typically used to determine molecular mass, molecular structure, morphology, thermal properties, and mechanical properties.

Emulsion polymerization

Emulsion polymerization is a type of radical polymerization that usually starts with an emulsion incorporating water, monomer, and surfactant. The most common type of emulsion polymerization is an oil-in-water emulsion, in which droplets of monomer (the oil) are emulsified (with surfactants) in a continuous phase of water. Water-soluble polymers, such as certain polyvinyl alcohols or hydroxyethyl celluloses, can also be used to act as emulsifiers/stabilizers. The name "emulsion polymerization" is a misnomer that arises from a historical misconception. Rather than occurring in emulsion droplets, polymerization takes place in the latex particles that form spontaneously in the first few minutes of the process. These latex particles are typically 100 nm in size, and are made of many individual polymer chains. The particles are stopped from coagulating with each other because each particle is surrounded by the surfactant ('soap'); the charge on the surfactant repels other particles electrostatically. When water-soluble polymers are used as stabilizers instead of soap, the repulsion between particles arises because these water-soluble polymers form a 'hairy layer' around a particle that repels other particles, because pushing particles together would involve compressing these chains.

Emulsion polymerization is used to manufacture several commercially important polymers. Many of these polymers are used as solid materials and must be isolated from the aqueous dispersion after polymerization. In other cases the dispersion itself is the end product. A dispersion resulting from emulsion polymerization is often called a latex (especially if derived from asynthetic rubber) or an emulsion (even though "emulsion" strictly speaking refers to a dispersion of an immiscible liquid in water). These emulsions find applications in adhesives, paints, paper coating and textile coatings. They are often preferred over solvent-based products in these applications due to the absence of VOCs (Volatile Organic Compounds) in them.

Techniques of polymerization

Advantages of emulsion polymerization include:

High molecular weight polymers can be made at fast polymerization rates. By contrast, in bulk and solution free radical polymerization, there is a tradeoff between molecular weight and polymerization rate.

The continuous water phase is an excellent conductor of heat, enabling fast polymerization rates without loss of temperature control.

Since polymer molecules are contained within the particles, the viscosity of the reaction medium remains close to that of water and is not dependent on molecular weight.

The final product can be used as is and does not generally need to be altered or processed.

Disadvantages of emulsion polymerization include:

Surfactants and other polymerization adjuvants remain in the polymer or are difficult to remove

For dry (isolated) polymers, water removal is an energy-intensive process

Emulsion polymerizations are usually designed to operate at high conversion of monomer to polymer. This can result in significant chain transfer to polymer.

It cannot be used for condensation, ionic or Ziegler-Natta polymerization, although some exceptions are known.

SOLUTION POLYMERIZATION

Solution polymerization is a method of industrial polymerization. In this procedure, a monomer is dissolved in a non-reactive solvent that contains a catalyst.

The reaction results in a polymer which is also soluble in the chosen solvent. Heat released by the reaction is absorbed by the solvent, and so the reaction rate is reduced. Moreover, the viscosity of the reaction mixture is reduced, not allowing auto acceleration at high monomer concentrations. Once the maximum or desired conversion is reached, excess solvent has to be removed in order to obtain the pure polymer. Hence, solution polymerization is mainly used for applications where the presence of a solvent is desired anyway, as is the case for varnish and adhesives. It is not useful for the production of dry polymers because of the difficulty of complete solvent removal.

This process is one of two used in the production of sodium polyacrylate, a super absorbent polymer used in disposable diapers.

Notable polymers produced using this method are polyacrylonitrile (PAN) and polyacrylic acid (PAA).

Solution polymerization is used to create polymers and copolymers by dissolving a monomer and a catalyst in a non-reactive solvent. During this process, the solvent liquid absorbs the heat generated by the chemical reaction which controls the reaction rate. The liquid solvent used in the solution polymerization procedure usually remains a solvent for the resulting polymer or copolymer. This process is only suitable for the creation of wet polymer types, as the removal of excess solvent is difficult. While the removal of the excess solvent is possible using distillation, it is usually not considered economically possible in an industrial situation.

The process of solution polymerization offers a few advantages as well as one major disadvantage. The advantages include precise control of the chemical reaction, control of the resulting heat and viscosity, and control over the auto acceleration of the process. The disadvantage of the process is the difficulty involved in the removal of excess solvent from the finished polymer.

The solvents used in the solution polymerization procedure must be chosen carefully. A solvent that is non-reactive to the monomer is essential to the process. If a reactive solvent is used, dangerous chain reaction processes or other undesirable effects can occur as a result of auto acceleration. Auto acceleration is a reaction that occurs when the heat produced by polymerization does not dissipate quickly enough by the solvent. As the heat builds up, the viscosity of the solution increases, causing the polymerization process to accelerate beyond safe control.

Industrial solution polymerization processes are commonly used to produce polymers with special characteristics. The chemical reaction that takes place between the monomer and catalyst can lend unique properties to the end product. One example of this type of polymer is sodium polyacrylate, the exceptionally absorbent polymer used in disposable diapers.

While this process is not generally feasible for dry polymers, it works well for wet polymer types. The process of industrial solution polymerization is used to create polymers and copolymers that can be used in their solution form. Examples of this usage include industrial glues and surface coatings.

Synthetic elastomers can also be produced using the solution polymerization process. This method produces a more precise polymer than emulsion polymerization methods. By controlling the addition of refined monomers to the catalyst-solvent solution, the resulting polymers can be carefully designed for specific properties. These synthetic elastomers are commonly found in products such as latex gloves, neoprene wetsuits and floor covering materials.

Suspension Polymerization

Suspension polymerization is used only in free radical type processes. The monomer is mechanically dispersed in a media, usually water. There are cases where an organic media is used in which neither the polymer nor the monomer is soluble in the organic media.

styrene- a liquid

vinyl chloride- a gas at room temperature

For suspension polymerization, there are two phases, water and organic, and the starting point may be 10 parts of the former, and 1 part of the latter. The initiator used can be water soluble or organic soluble [benzoyl peroxide, AIBN, or (NH4)2(SxO4)y.] Usually the initiator is organic soluble.

There are two separate phases throughout the whole process.

The droplets must be kept far apart. This requires agitation: consistent, efficient, and controlled. A suspending agent can be used. Poly(vinyl alcohol) dissolved in the aqueous phase is a typical suspending agent.

The rate of suspension polymerization is similar to the rate of bulk polymerization, but the heat transfer is much better. For suspension polymerization, initiation, propagation, and termination take place inside the droplet.

examples include the polymerization of methyl methacrylate and vinyl chloride.

The solution polymerization of butadiene, isobutylene and isoprene require a pressure system.

the media to monomer ratio is 10:1. Is this wt/vol, wt/wt, or vol/vol?? In an industrial setting it may be easier to combine chemical feedstocks on a per volume basis. Ask Dr. Stoffer.

Particle size is affect by the following four factors:

stirring rate

ratio of reactants

suspension agent

temperature

The plant operator must control temperature, and the particle size (of the growing polymer mass in the bubble.) If the particle size gets to large, the particle will absorb too much heat. This probably relates to the idea that as you increase the volume of a sphere, the ratio of surface area to volume decreases, and this ratio relates to heat transfer. Particle size may be 0.01 to 0.5 cm, or as low as 1 micron.

A suspension agent is a material that gives a surface activation that keeps droplets from become larger (droplets coming together to form larger droplets is called coalescence.)

I think the term coalescence is also used for when the polymer molecules in a solution come together as the solvent evaporates, an example of which is the film formation that occurs for a paint system.

Suspension polymerization is similar to bulk polyerization, and it could be considered "bulk polymerization within a droplet." The speed at which the reaction takes place for a given temperature is the same, and just as for bulk polymerization, the kinetics or rates are proportional to monomer concentration.

There is efficient heat transfer with speed or bulk process.

Recovery of product by mechanical separation (i.e, filtration, washing) is a lot cheaper than thermo distillation. (this sentence sort of doesn't make sense.)

The properties of the polymer are similar to those of the same polymer made by a bulk polymerization.

Limitations of suspension polymerization–

It only applies to free radical process.

Ionic catalysts don't work because they compete with water.

Agitation is critical because as the viscosity within the bead rises, the reaction rate increases suddenly (Tromsdoiff effect.) This leads to a surge in heat generation which does not usually occur in solution or emulsion polymerization.

Importance of Polymerization

It is necessary for every student to learn and understand each and every topic in detail, which are included in the syllabus of chemistry. And the same goes for the topic of Polymerisation, the topic is very important in the subject of Chemistry, and hence students need to have a good understanding of the meaning of Polymerisation, the types of Polymerisations, as well as degrees and techniques of polymerization.

But it is just as important for the students to have an explanation that is easy to understand and comprehend, otherwise what happens is a student may find the topic difficult and may give up on it as well. And hence, to avoid such a situation for the students, Vedantu has brought the complete Explanation of the Polymerisation, along with its types, techniques, and Degrees. So that the students can understand the topic easily without any problem.

Vedantu made it possible because the explanation that Vedantu provides for the Polymerisation, along with its types, techniques, and Degrees, is presented in a manner that is easy for eyes and easy for brains. That is to say, the explanation is well formatted so that students can understand each of the sub-topic of the main topic of Polymerisation. And easy for the brain, because the explanation of Polymerisation, along with its types, techniques, and Degrees is written in a simple and lucid language.

Before discussing polymerization, let us first have a quick look at the "Monomer Molecules" and the "Polymer". So, "mono" means one, and "mer" means part. Monomer molecules are those molecules, which can react with the other molecules and, therefore, manage to form a very large molecule. Now coming to the second word which is a polymer. The word polymer itself means Many (poly) parts (Mer), and hence the macromolecules or a substance or a material that consists of very large molecules, are called polymers. Now that the meaning of "Monomer Molecules" and the "polymer" is clear, understanding the process of polymerisation becomes a little easier to understand. So, when the monomer Molecules react together in a chemical reaction and form a polymer chain or three-dimensional network, then this process is called Polymerisation.

Types of Polymerisation

There are many ways in which Polymerisation is formed, a few of which are discussed below.

Step-growth Polymerisation: When the pairs of reactants of any length, fuse or combine at each step in order to form a longer polymer molecule, then this particular way of forming the polymerization is regarded as the step-growth polymerization. In this process of polymerization, the average molar mass increases gradually but slowly, while the long chains take time to be formed.

Chain-growth Polymerisation: When the step of chain extension adds to the monomer in a growing chain, with an active center such as cation, or free radical, then such formation of the polymerization, is regarded as a Chain-growth polymerization because here a chain reaction takes place. The unsaturated monomers, especially those which contain the double bonds of carbon-carbon, are all linked together in the chain-growth polymerization. While due to the formation of a new sigma bond the pi bond is lost.

Condensation Polymerization

Condensation polymers are formed by a reaction between two different bifunctional or tri-functional monomer units. These monomeric units undergo a condensation reaction to form polymers and small molecules of simple compounds like HCl (Hydrogen chloride), H2O (Water), alcohol, etc. are eliminated. Silicon, Bakelite, Nylon, etc. are some of the examples of condensation polymers.

Addition Polymerization

In case of addition polymerisation, double or triple bond monomers are repeatedly added to form a polymer. In this type of reaction, no by-products are formed. For instance, propene (C3H6) helps in the formation of poly-propene ((C3H6)n) and polythene ((C2H4)n) is formed by the addition reaction of ethene (C2H4) molecules.

This can be Further Divided into Two Categories –

Copolymers – Those polymers which are formed due to the addition reaction held between two different kinds of monomers are known as copolymers. Buna-N and Buna-S are examples of homopolymers.

Homopolymers – Those polymers which are formed due to the addition reaction held between the same kinds of monomers are known as Homopolymers. Polythene made from a single kind of monomer ethene is an example of a homopolymer.

Photopolymerisation reactions are also chain-growth polymerization. When the ultraviolet light is absorbed, then it initiates the photopolymerization reaction. The absorbed light can either be direct or indirect, if it is direct then it is due to the reactant monomer, and when it is not direct then it is due to the photosensitizers, which absorbs the light, that is to say, ultraviolet light, and then transfers this energy into the monomer.

The only difference between the photopolymerization and the ordinary thermal polymerization is that of the initiation because the step of initiation is different in both the polymerization.

Glass transition temperature (Tg)

The glass–liquid transition, or glass transition, is the gradual and reversible transition in amorphous materials (or in amorphous regions within semicrystalline materials) from a hard and relatively brittle "glassy" state into a viscous or rubbery state as the temperature is increased. An amorphous solid that exhibits a glass transition is called a glass. The reverse transition, achieved by supercooling a viscous liquid into the glass state, is called vitrification.

The glass-transition temperature Tg of a material characterizes the range of temperatures over which this glass transition occurs. It is always lower than the melting temperature, Tm, of the crystalline state of the material, if one exists.

Hard plastics like polystyrene and poly(methyl methacrylate) are used well below their glass transition temperatures, i.e., when they are in their glassy state. Their Tg values are both at around 100 °C (212 °F). Rubber elastomers like polyisoprene and polyisobutylene are used above their Tg, that is, in the rubbery state, where they are soft and flexible; crosslinking prevents free flow of their molecules, thus endowing rubber with a set shape at room temperature (as opposed to a viscous liquid).

Despite the change in the physical properties of a material through its glass transition, the transition is not considered a phase transition; rather it is a phenomenon extending over a range of temperature and defined by one of several conventions. Such conventions include a constant cooling rate (20 kelvins per minute (36 °F/min)) and a viscosity threshold of 1012 Pa·s, among others. Upon cooling or heating through this glass-transition range, the material also exhibits a smooth step in the thermal-expansion coefficient and in the specific heat, with the location of these effects again being dependent on the history of the material. The question of whether some phase transition underlies the glass transition is a matter of continuing research. [when?]

The glass transition of a liquid to a solid-like state may occur with either cooling or compression. The transition comprises a smooth increase in the viscosity of a material by as much as 17 orders of magnitude within a temperature range of 500 K without any pronounced change in material structure. The consequence of this dramatic increase is a glass exhibiting solid-like mechanical properties on the timescale of practical observation. This transition is in contrast to the freezing or crystallization transition, which is a first-order phase transition in the Ehrenfest classification and involves discontinuities in thermodynamic and dynamic properties such as volume, energy, and viscosity. In many materials that normally undergo a freezing transition, rapid cooling will avoid this phase transition and instead result in a glass transition at some lower temperature. Other materials, such as many polymers, lack a well defined crystalline state and easily form glasses, even upon very slow cooling or compression. The tendency for a material to form a glass while quenched is called glass forming ability. This ability depends on the composition of the material and can be predicted by the rigidity theory.

Below the transition temperature range, the glassy structure does not relax in accordance with the cooling rate used. The expansion coefficient for the glassy state is roughly equivalent to that of the crystalline solid. If slower cooling rates are used, the increased time for structural relaxation (or intermolecular rearrangement) to occur may result in a higher density glass product. Similarly, by annealing (and thus allowing for slow structural relaxation) the glass structure in time approaches an equilibrium density corresponding to the supercooled liquid at this same temperature. Tg is located at the intersection between the cooling curve (volume versus temperature) for the glassy state and the supercooled liquid.

The configuration of the glass in this temperature range changes slowly with time towards the equilibrium structure. The principle of the minimization of the Gibbs free energy provides the thermodynamic driving force necessary for the eventual change. At somewhat higher temperatures than Tg, the structure corresponding to equilibrium at any temperature is achieved quite rapidly. In contrast, at considerably lower temperatures, the configuration of the glass remains sensibly stable over increasingly extended periods of time.

Thus, the liquid-glass transition is not a transition between states of thermodynamic equilibrium. It is widely believed that the true equilibrium state is always crystalline. Glass is believed to exist in a kinetically locked state, and its entropy, density, and so on, depend on the thermal history. Therefore, the glass transition is primarily a dynamic phenomenon. Time and temperature are interchangeable quantities (to some extent) when dealing with glasses, a fact often expressed in the time–temperature superposition principle. On cooling a liquid, internal degrees of freedom successively fall out of equilibrium. However, there is a longstanding debate whether there is an underlying second-order phase transition in the hypothetical limit of infinitely long relaxation times.

In a more recent model of glass transition, the glass transition temperature corresponds to the temperature at which the largest openings between the vibrating elements in the liquid matrix become smaller than the smallest cross-sections of the elements or parts of them when the temperature is decreasing. As a result of the fluctuating input of thermal energy into the liquid matrix, the harmonics of the oscillations are constantly disturbed and temporary cavities ("free volume") are created between the elements, the number and size of which depend on the temperature. The glass transition temperature Tg0 defined in this way is a fixed material constant of the disordered (non-crystalline) state that is dependent only on the pressure. As a result of the increasing inertia of the molecular matrix when approaching Tg0, the setting of the thermal equilibrium is successively delayed, so that the usual measuring methods for determining the glass transition temperature in principle deliver Tg values that are too high. In principle, the slower the temperature change rate is set during the measurement, the closer the measured Tg value Tg0 approaches. Techniques such as dynamic mechanical analysis can be used to measure the glass transition temperature.

TRANSITION TEMPERATURE TG

Refer to the figure on the bottom right plotting the heat capacity as a function of temperature. In this context, Tg is the temperature corresponding to point A on the curve.

Different operational definitions of the glass transition temperature Tg are in use, and several of them are endorsed as accepted scientific standards. Nevertheless, all definitions are arbitrary, and all yield different numeric results: at best, values of Tg for a given substance agree within a few kelvins. One definition refers to the viscosity, fixing Tg at a value of 1013 poise (or 1012 Pa·s). As evidenced experimentally, this value is close to the annealing point of many glasses.

In contrast to viscosity, the thermal expansion, heat capacity, shear modulus, and many other properties of inorganic glasses show a relatively sudden change at the glass transition temperature. Any such step or kink can be used to define Tg. To make this definition reproducible, the cooling or heating rate must be specified.

The most frequently used definition of Tg uses the energy release on heating in differential scanning calorimetry (DSC, see figure). Typically, the sample is first cooled with 10 K/min and then heated with that same speed.

Yet another definition of Tg uses the kink in dilatometry (a.k.a. thermal expansion): refer to the figure on the top right. Here, heating rates of 3–5 K/min (5.4–9.0 °F/min) are common. The linear sections below and above Tg are colored green. Tg is the temperature at the intersection of the red regression lines.

Summarized below are Tg values characteristic of certain classes of materials.

In specific materials

Silica, SiO2

Silica (the chemical compound SiO2) has a number of distinct crystalline forms in addition to the quartz structure. Nearly all of the crystalline forms involve tetrahedral SiO4 units linked together by shared vertices in different arrangements (stishovite, composed of linked SiO6 octahedra, is the main exception). Si-O bond lengths vary between the different crystal forms. For example, in a-quartz the bond length is 161 picometres (6.3×10-9 in), whereas in a-tridymite it ranges from 154–171 pm (6.1×10-9–6.7×10-9 in). The Si-O-Si bond angle also varies from 140° in a-tridymite to 144° in a-quartz to 180° in ß-tridymite. Any deviations from these standard parameters constitute microstructural differences or variations that represent an approach to an amorphous, vitreous or glassy solid. The transition temperature Tg in silicates is related to the energy required to break and re-form covalent bonds in an amorphous (or random network) lattice of covalent bonds. The Tg is clearly influenced by the chemistry of the glass. For example, addition of elements such as B, Na, K or Ca to a silica glass, which have a valency less than 4, helps in breaking up the network structure, thus reducing the Tg. Alternatively, P, which has a valency of 5, helps to reinforce an ordered lattice, and thus increases the Tg. Tg is directly proportional to bond strength, e.g. it depends on quasi-equilibrium thermodynamic parameters of the bonds e.g. on the enthalpy Hd and entropy Sd of configurons – broken bonds: Tg = Hd / [Sd + R ln[(1 - fc)/ fc] where R is the gas constant and fc is the percolation threshold. For strong melts such as SiO2 the percolation threshold in the above equation is the universal Scher–Zallen critical density in the 3-D space e.g. fc = 0.15, however for fragile materials the percolation thresholds are material-dependent and fc « 1. The enthalpy Hd and the entropy Sd of configurons – broken bonds can be found from available experimental data on viscosity.

Polymers

In polymers the glass transition temperature, Tg, is often expressed as the temperature at which the Gibbs free energy is such that the activation energy for the cooperative movement of 50 or so elements of the polymer is exceeded . This allows molecular chains to slide past each other when a force is applied. From this definition, we can see that the introduction of relatively stiff chemical groups (such as benzene rings) will interfere with the flowing process and hence increase Tg. The stiffness of thermoplastics decreases due to this effect (see figure.) When the glass temperature has been reached, the stiffness stays the same for a while, i.e., at or near E2, until the temperature exceeds Tm, and the material melts. This region is called the rubber plateau.

Coming from the low-temperature side, the shear modulus drops by many orders of magnitude at the glass transition temperature Tg. A molecular-level mathematical relation for the temperature-dependent shear modulus of the polymer glass on approaching Tg from below has been developed by Alessio Zaccone and Eugene Terentjev. Even though the shear modulus does not really drop to zero (it drops down to the much lower value of the rubber plateau), upon setting the shear modulus to zero in the Zaccone–Terentjev formula, an expression for Tg is obtained which recovers the Flory–Fox equation, and also shows that Tg is inversely proportional to the thermal expansion coefficient in the glass state. This procedure provides yet another operational protocol to define the Tg of polymer glasses by identifying it with the temperature at which the shear modulus drops by many orders of magnitude down to the rubbery plateau.

In ironing, a fabric is heated through this transition so that the polymer chains become mobile. The weight of the iron then imposes a preferred orientation. Tg can be significantly decreased by addition of plasticizers into the polymer matrix. Smaller molecules of plasticizer embed themselves between the polymer chains, increasing the spacing and free volume, and allowing them to move past one another even at lower temperatures. Addition of plasticizer can effectively take control over polymer chain dynamics and dominate the amounts of the associated free volume so that the increased mobility of polymer ends is not apparent. The addition of nonreactive side groups to a polymer can also make the chains stand off from one another, reducing Tg. If a plastic with some desirable properties has a Tg that is too high, it can sometimes be combined with another in a copolymer or composite material with a Tg below the temperature of intended use. Note that some plastics are used at high temperatures, e.g., in automobile engines, and others at low temperatures.

Mechanics of vitrification

Molecular motion in condensed matter can be represented by a Fourier series whose physical interpretation consists of a superposition of longitudinal and transverse waves of atomic displacement with varying directions and wavelengths. In monatomic systems, these waves are called density fluctuations. (In polyatomic systems, they may also include compositional fluctuations.)

Thus, thermal motion in liquids can be decomposed into elementary longitudinal vibrations (or acoustic phonons) while transverse vibrations (or shear waves) were originally described only in elastic solids exhibiting the highly ordered crystalline state of matter. In other words, simple liquids cannot support an applied force in the form of a shearing stress, and will yield mechanically via macroscopic plastic deformation (or viscous flow). Furthermore, the fact that a solid deforms locally while retaining its rigidity – while a liquid yields to macroscopic viscous flow in response to the application of an applied shearing force – is accepted by many as the mechanical distinction between the two.

The inadequacies of this conclusion, however, were pointed out by Frenkel in his revision of the kinetic theory of solids and the theory of elasticity in liquids. This revision follows directly from the continuous characteristic of the viscoelastic crossover from the liquid state into the solid one when the transition is not accompanied by crystallization—ergo the supercooled viscous liquid. Thus we see the intimate correlation between transverse acoustic phonons (or shear waves) and the onset of rigidity upon vitrification, as described by Bartenev in his mechanical description of the vitrification process. This concept leads to defining the glass transition in terms of the vanishing or significant lowering of the low-frequency shear modulus, as shown quantitatively in the work of Zaccone and Terentjev on the example of polymer glass. In fact, the shoving model stipulates that the activation energy of the relaxation time is proportional to the high-frequency plateau shear modulus, a quantity that increases upon cooling thus explaining the ubiquitous non-Arrhenius temperature dependence of the relaxation time in glass-forming liquids.

The velocities of longitudinal acoustic phonons in condensed matter are directly responsible for the thermal conductivity that levels out temperature differentials between compressed and expanded volume elements. Kittel proposed that the behavior of glasses is interpreted in terms of an approximately constant "mean free path" for lattice phonons, and that the value of the mean free path is of the order of magnitude of the scale of disorder in the molecular structure of a liquid or solid. The thermal phonon mean free paths or relaxation lengths of a number of glass formers have been plotted versus the glass transition temperature, indicating a linear relationship between the two. This has suggested a new criterion for glass formation based on the value of the phonon mean free path.

It has often been suggested that heat transport in dielectric solids occurs through elastic vibrations of the lattice, and that this transport is limited by elastic scattering of acoustic phonons by lattice defects (e.g. randomly spaced vacancies). These predictions were confirmed by experiments on commercial glasses and glass ceramics, where mean free paths were apparently limited by "internal boundary scattering" to length scales of 10–100 micrometres (0.00039–0.00394 in). The relationship between these transverse waves and the mechanism of vitrification has been described by several authors who proposed that the onset of correlations between such phonons results in an orientational ordering or "freezing" of local shear stresses in glass-forming liquids, thus yielding the glass transition.

Electronic structure

The influence of thermal phonons and their interaction with electronic structure is a topic that was appropriately introduced in a discussion of the resistance of liquid metals. Lindemann's theory of melting is referenced, and it is suggested that the drop in conductivity in going from the crystalline to the liquid state is due to the increased scattering of conduction electrons as a result of the increased amplitude of atomic vibration. Such theories of localization have been applied to transport in metallic glasses, where the mean free path of the electrons is very small (on the order of the interatomic spacing).

The formation of a non-crystalline form of a gold-silicon alloy by the method of splat quenching from the melt led to further considerations of the influence of electronic structure on glass forming ability, based on the properties of the metallic bond.

Other work indicates that the mobility of localized electrons is enhanced by the presence of dynamic phonon modes. One claim against such a model is that if chemical bonds are important, the nearly free electron models should not be applicable. However, if the model includes the buildup of a charge distribution between all pairs of atoms just like a chemical bond (e.g., silicon, when a band is just filled with electrons) then it should apply to solids.

Thus, if the electrical conductivity is low, the mean free path of the electrons is very short. The electrons will only be sensitive to the short-range order in the glass since they do not get a chance to scatter from atoms spaced at large distances. Since the short-range order is similar in glasses and crystals, the electronic energies should be similar in these two states. For alloys with lower resistivity and longer electronic mean free paths, the electrons could begin to sense[dubious – discuss] that there is disorder in the glass, and this would raise their energies and destabilize the glass with respect to crystallization. Thus, the glass formation tendencies of certain alloys may therefore be due in part to the fact that the electron mean free paths are very short, so that only the short-range order is ever important for the energy of the electrons.

It has also been argued that glass formation in metallic systems is related to the "softness" of the interaction potential between unlike atoms. Some authors, emphasizing the strong similarities between the local structure of the glass and the corresponding crystal, suggest that chemical bonding helps to stabilize the amorphous structure.

Other authors have suggested that the electronic structure yields its influence on glass formation through the directional properties of bonds. Non-crystallinity is thus favored in elements with a large number of polymorphic forms and a high degree of bonding anisotropy. Crystallization becomes more unlikely as bonding anisotropy is increased from isotropic metallic to anisotropic metallic to covalent bonding, thus suggesting a relationship between the group number in the periodic table and the glass forming ability in elemental solids.

3

PRINCIPLES OF CHROMATOGRAPHY

In chemical analysis, chromatography is a laboratory technique for the separation of a mixture into its components. The mixture is dissolved in a fluid solvent (gas or liquid) called the mobile phase, which carries it through a system (a column, a capillary tube, a plate, or a sheet) on which a material called the stationary phase is fixed. Because the different constituents of the mixture tend to have different affinities for the stationary phase and are retained for different lengths of time depending on their interactions with its surface sites, the constituents travel at different apparent velocities in the mobile fluid, causing them to separate. The separation is based on the differential partitioning between the mobile and the stationary phases. Subtle differences in a compound's partition coefficient result in differential retention on the stationary phase and thus affect the separation.

Chromatography may be preparative or analytical. The purpose of preparative chromatography is to separate the components of a mixture for later use, and is thus a form of purification. This process is associated with higher costs due to its mode of production. Analytical chromatography is done normally with smaller amounts of material and is for establishing the presence or measuring the relative proportions of analytes in a mixture. The two types are not mutually exclusive.

'Chromatography' is an analytical technique commonly used for separating a mixture of chemical substances into its individual components, so that the individual components can be thoroughly analyzed. There are many types of chromatography e.g., liquid chromatography, gas chromatography, ion-exchange chromatography, affinity chromatography, but all of these employ the same basic principles.

FROM ART TO SCIENCE

Chromatography was initially used by artists, color theorists and artisans hoping to perfect industrial dyes for textiles. With time, it also spawned a unique branch of chemistry, and with it, the techniques used today to understand and purify mixtures.

In modern laboratories, the color aspect is no longer relevant, but the same principles apply. By dissolving a mixture of interest in a mobile phase and transporting it through a stationary phase, the components of the mixture can be separated from one another based on their different speeds of travel.

By altering the mobile phase, the stationary phase, and/or the factor determining speed of travel, a wide variety of chromatographic methods have been created, each serving a different purpose and ideal for different mixtures. Some of the most common forms of chromatography are as follows.

In gas chromatography, the mixture of interest is vaporized and carried through a stationary phase (usually a metal or glass separation column) with an inert gas, usually nitrogen or helium. Larger molecules in the mixture take longer to pass through the column and reach the detector at the far end.

In liquid chromatography, the mixture of interest is dissolved in a liquid and passed through a solid stationary phase, which is often made of a silica material. Several varieties of liquid chromatography exist, depending on the relative polarities of the mobile and stationary phases (normal-phase versus reverse-phase) and whether the mobile phase is pressurized (high-performance).

In thin-layer chromatography (TLC), the stationary phase is a thin layer of solid material, usually silica-based, and the mobile phase is a liquid in which the mixture of interest is dissolved. Thin-layer chromatography comes with the advantage of photographing well, making its output easy to digitize.

Ion exchange chromatography separates the components of a mixture based on their charge, in addition to or instead of their size. In essence, positively (cations) or negatively (anions) charged ions are separated using different stationary phases and different pH mobile phases.

Chromatography can be used as an analytical tool, feeding its output into a detector that reads the contents of the mixture. It can also be used as a purification tool, separating the components of a mixture for use in other experiments or procedures. Typically, analytical chromatography uses a much smaller quantity of material than chromatography meant to purify a mixture or extract specific components from it.

For example, solid-phase extraction is a kind of liquid chromatography in which different mobile phases are used in sequence to separate out different components of a mixture trapped in a solid phase. Chromatography as a purification technique has major roles in petrochemical and other organic chemistry laboratories, where it can be one of the more cost-effective ways to remove impurities from organic solutions, particularly if the components of the mixture are heat-sensitive.

Flexibility

The principles of chromatography also appear in other laboratory techniques. Gel electrophoresis sorts nucleic acids and proteins based on their size, drawing them through the gel via an electric field. This technique is, in effect, a kind of chromatography. Similarly, distillation sorts the components of a mixture by their boiling and condensation points, with the apparatus itself being a sort of stationary phase.

Because its core principle is so simple, chromatography leaves room for substantial refinement. This has led to a variety of more specialized chromatographic techniques, such as two-dimensional chromatography for using two different chromatography methods at once, pyrolysis gas chromatography, used as part of mass spectrometry, and chiral chromatography, which is used to separate stereoisomers that other methods cannot distinguish.

Chromatography is a simple and exceedingly flexible principle, that will continue to spawn new variations and new implementations into the foreseeable future.

Chromatography terms

- The analyte is the substance to be separated during chromatography.
- Analytical chromatography is used to determine the existence and possibly also the concentration of analyte(s) in a sample.
- A bonded phase is a stationary phase that is covalently bonded to the support particles or to the inside wall of the column tubing.
- A chromatogram is the visual output of the chromatograph. In the case of an optimal separation, different peaks or patterns on the chromatogram correspond to different components of the separated mixture.

Plotted on the x-axis is the retention time and plotted on the y-axis a signal (for example obtained by a spectrophotometer, mass spectrometer or a variety of other detectors) corresponding to the response created by the analytes exiting the system. In the case of an optimal system the signal is proportional to the concentration of the specific analyte separated:

- A chromatograph is equipment that enables a sophisticated separation e.g. gas chromatographic or liquid chromatographic separation.
- Chromatography is a physical method of separation that distributes components to separate between two phases, one stationary (stationary phase), while the other (the mobile phase) moves in a definite direction.
- The eluate is the mobile phase leaving the column.
- The eluent is the solvent that carries the analyte.
- An eluotropic series is a list of solvents ranked according to their eluting power.
- An immobilized phase is a stationary phase that is immobilized on the support particles, or on the inner wall of the column tubing.
- The mobile phase is the phase that moves in a definite direction. It may be a liquid (LC and Capillary Electrochromatography (CEC)), a gas (GC), or a supercritical fluid (supercritical-fluid chromatography, SFC). The mobile phase consists of the sample being separated/analyzed and the solvent that moves the sample through the column. In the case of HPLC the mobile phase consists of a non-polar solvent(s) such as hexane in normal phase or polar solvents in reverse phase chromotagraphy and the sample being separated. The mobile phase moves through the chromatography column (the stationary phase) where the sample interacts with the stationary phase and is separated.
- Preparative chromatography is used to purify sufficient quantities of a substance for further use, rather than analysis.
- The retention time is the characteristic time it takes for a particular analyte to pass through the system (from the column inlet to the detector) under set conditions. See also: Kovats' retention index
- The sample is the matter analyzed in chromatography. It may consist of a single component or it may be a mixture of components. When the sample is treated in the course of an analysis, the phase or the phases containing the analytes of interest is/are referred to as the sample whereas everything out of interest separated from the sample before or in the course of the analysis is referred to as waste.
- The solute refers to the sample components in partition chromatography.
- The solvent refers to any substance capable of solubilizing another substance, and especially the liquid mobile phase in liquid chromatography.

- The stationary phase is the substance fixed in place for the chromatography procedure. Examples include the silica layer in thin layer chromatography

Chromatography is based on the concept of partition coefficient. Any solute partitions between two immiscible solvents. When we make one solvent immobile (by adsorption on a solid support matrix) and another mobile it results in most common applications of chromatography. If matrix support is polar (e.g. paper, silica etc.) it is forward phase chromatography, and if it is non polar (C-18) it is reverse phase.

TECHNIQUES BY CHROMATOGRAPHIC BED SHAPE

Column chromatography

Column chromatography is a separation technique in which the stationary bed is within a tube. The particles of the solid stationary phase or the support coated with a liquid stationary phase may fill the whole inside volume of the tube (packed column) or be concentrated on or along the inside tube wall leaving an open, unrestricted path for the mobile phase in the middle part of the tube (open tubular column). Differences in rates of movement through the medium are calculated to different retention times of the sample.

Column chromatography in chemistry is a method used to purify individual chemical compounds from mixtures of compounds. It is often used for preparative applications on scales from micrograms up to kilograms. The main advantage of column chromatography is the relatively low cost and disposability of the stationary phase used in the process. The latter prevents cross-contamination and stationary phase degradation due to recycling.

The classical preparative chromatography column, is a glass tube with a diameter from 5 mm to 50 mm and a height of 5 cm to 1 m with a tap and some kind of a filter (a glass frit or glass wool plug – to prevent the loss of the stationary phase) at the bottom. Two methods are generally used to prepare a column: the dry method, and the wet method.

- For the dry method, the column is first filled with dry stationary phase powder, followed by the addition of mobile phase, which is flushed through the column until it is completely wet, and from this point is never allowed to run dry.

- For the wet method, a slurry is prepared of the eluent with the stationary phase powder and then carefully poured into the column. Care must be taken to avoid air bubbles. A solution of the organic material is pipetted on top of the stationary phase. This layer is usually topped with a small layer of sand or with cotton or glass wool to protect the shape of the organic layer from the velocity of newly added eluent. Eluent is slowly passed through the column to advance the organic material. Often a spherical eluent reservoir or an eluent-filled and stoppered separating funnel is put on top of the column.

The individual components are retained by the stationary phase differently and separate from each other while they are running at different speeds through the column with the eluent. At the end of the column they elute one at a time. During the entire chromatography process the eluent is collected in a series of fractions. The composition of the eluent flow can be monitored and each fraction is analyzed for dissolved compounds, e.g. by analytical chromatography, UV absorption, or fluorescence. Colored compounds (or fluorescent compounds with the aid of an UV lamp) can be seen through the glass wall as moving bands.

In 1978, W. C. Still introduced a modified version of column chromatography called flash column chromatography (flash). The technique is very similar to the traditional column chromatography, except for that the solvent is driven through the column by applying positive pressure. This allowed most separations to be performed in less than 20 minutes, with improved separations compared to the old method. Modern flash chromatography systems are sold as pre-packed plastic cartridges, and the solvent is pumped through the cartridge. Systems may also be linked with detectors and fraction collectors providing automation. The introduction of gradient pumps resulted in quicker separations and less solvent usage.

In expanded bed adsorption, a fluidized bed is used, rather than a solid phase made by a packed bed. This allows omission of initial clearing steps such as centrifugation and filtration, for culture broths or slurries of broken cells.

Phosphocellulose chromatography utilizes the binding affinity of many DNA-binding proteins for phosphocellulose. The stronger a protein's interaction with DNA, the higher the salt concentration needed to elute that protein.

PLANAR CHROMATOGRAPHY

Planar chromatography is a separation technique in which the stationary phase is present as or on a plane. The plane can be a paper, serving as such or impregnated by a substance as the stationary bed (paper chromatography) or a layer of solid particles spread on a support such as a glass plate (thin layer chromatography). Different compounds in the sample mixture travel different distances according to how strongly they interact with the stationary phase as compared to the mobile phase. The specific Retention factor (Rf) of each chemical can be used to aid in the identification of an unknown substance.

Paper chromatography

Paper chromatography is an analytical method technique for separating and identifying mixtures that are or can be colored, especially pigments. This can also be used in secondary or primary colors in ink experiments. This method has been largely replaced by thin layer chromatography, however it is still a powerful teaching tool. Double-way paper chromatography, also called two-dimensional chromatography, involves using two solvents and rotating the paper 90° in between. This is useful for separating complex mixtures of similar compounds, for example, amino acids. Paper chromatography is a technique that involves placing a small dot or line of sample solution onto a strip of chromatography paper. The paper is placed in a jar containing a shallow layer of solvent and sealed. As the solvent rises through the paper, it meets the sample mixture, which starts to travel up the paper with the solvent. This paper is made of cellulose, a polar substance, and the compounds within the mixture travel farther if they are non-polar. More polar substances bond with the cellulose paper more quickly, and therefore do not travel as far.

Thin layer chromatography

Thin layer chromatography (TLC) is a chromatography technique used to separate mixtures. Thin layer chromatography is performed on a sheet of glass, plastic, or aluminium foil, which is coated with a thin layer of adsorbent material, usually silica gel, aluminium oxide, or cellulose (blotter paper). This layer of adsorbent is known as the stationary phase.

After the sample has been applied on the plate, a solvent or solvent mixture (known as the mobile phase) is drawn up the plate via capillary action. Because different analytes ascend the TLC plate at different rates, separation is achieved.

Thin layer chromatography can be used to monitor the progress of a reaction, identify compounds present in a given mixture, and determine the purity of a substance. Specific examples of these applications include: analyzing ceramides and fatty acids, detection of pesticides or insecticides in food and water, analyzing the dye composition of fibers in forensics, assaying the radiochemical purity of radiopharmaceuticals, or identification of medicinal plants and their constituents

A number of enhancements can be made to the original method to automate the different steps, to increase the resolution achieved with TLC and to allow more accurate quantization. This method is referred to as HPTLC, or "high performance TLC".

Thin layer chromatography (TLC) is a widely employed laboratory technique and is similar to paper chromatography. However, instead of using a stationary phase of paper, it involves a stationary phase of a thin layer of adsorbent like silica gel, alumina, or cellulose on a flat, inert substrate. Compared to paper, it has the advantage of faster runs, better separations, and the choice between different adsorbents. For even better resolution and to allow for quantification, high-performance TLC can be used.

Displacement chromatography

Displacement chromatography is a chromatography technique in which a sample is placed onto the head of the column[n 1] and is then displaced by a solute that is more strongly sorbed than the components of the original mixture. The result is that the components are resolved into consecutive "rectangular" zones of highly concentrated pure substances rather than solvent-separated "peaks". It is primarily a preparative technique; higher product concentration, higher purity, and increased throughput may be obtained compared to other modes of chromatography.

The basic principle of displacement chromatography is: A molecule with a high affinity for the chromatography matrix (the displacer) competes effectively for binding sites, and thus displace all molecules with lesser affinities. There are distinct differences between displacement and elution chromatography. In elution mode, substances typically emerge from a column in narrow, Gaussian peaks. Wide separation of peaks, preferably to baseline, is desired for maximum purification. The speed at which any component of a mixture travels down the column in elution mode depends on many factors. But for two substances to travel at different speeds, and thereby be resolved, there must be substantial differences in some interaction between the biomolecules and the chromatography matrix. Operating parameters are adjusted to maximize the effect of this difference. In many cases, baseline separation of the peaks can be achieved only with gradient elution and low column loadings. Thus, two drawbacks to elution mode chromatography, especially at the preparative scale, are operational complexity, due to gradient solvent pumping, and low throughput, due to low column loadings. Displacement chromatography has advantages over elution chromatography in that components are resolved into consecutive zones of pure substances rather than "peaks". Because the process takes advantage of the nonlinearity of the isotherms, a larger column feed can be separated on a given column with the purified components recovered at significantly higher concentrations.

Gas chromatography

Gas chromatography (GC), also sometimes known as gas-liquid chromatography, (GLC), is a separation technique in which the mobile phase is a gas. Gas chromatography is always carried out in a column, which is typically "packed" or "capillary".

Gas chromatography (GC), is a common type of chromatography used in analytical chemistry for separating and analyzing compounds that can be vaporized without decomposition. Typical uses of GC include testing the purity of a particular substance, or separating the different components of a mixture (the relative amounts of such components can also be determined). In some situations, GC may help in identifying a compound. In preparative chromatography, GC can be used to prepare pure compounds from a mixture.

In gas chromatography, the mobile phase (or "moving phase") is a carrier gas, usually an inert gas such as helium or an unreactive gas such as nitrogen. The stationary phase is a microscopic layer of liquid or polymer on an inert solid support, inside a piece of glass or metal tubing called a column (a homage to the fractionating column used in distillation). The instrument used to perform gas chromatography is called a gas chromatograph (or "aerograph", "gas separator").

The gaseous compounds being analyzed interact with the walls of the column, which is coated with a stationary phase. This causes each compound to elute at a different time, known as the retention time of the compound. The comparison of retention times is what gives GC its analytical usefulness.

Gas chromatography is in principle similar to column chromatography (as well as other forms of chromatography, such as HPLC, TLC), but has several notable differences. Firstly, the process of separating the compounds in a mixture is carried out between a liquid stationary phase and a gas mobile phase, whereas in column chromatography the stationary phase is a solid and the mobile phase is a liquid. (Hence the full name of the procedure is "Gas–liquid chromatography", referring to the mobile and stationary phases, respectively.) Secondly, the column through which the gas phase passes is located in an oven where the temperature of the gas can be controlled, whereas column chromatography (typically) has no such temperature control. Thirdly, the concentration of a compound in the gas phase is solely a function of the vapor pressure of the gas.

Gas chromatography is also similar to fractional distillation, since both processes separate the components of a mixture primarily based on boiling point (or vapor pressure) differences. However, fractional distillation is typically used to separate components of a mixture on a large scale, whereas GC can be used on a much smaller scale (i.e. microscale).

Gas chromatography is based on a partition equilibrium of analyte between a solid stationary phase (often a liquid silicone-based material) and a mobile gas (most often helium). The stationary phase is adhered to the inside of a small-diameter glass tube (a capillary column) or a solid matrix inside a larger metal tube (a packed column). It is widely used in analytical chemistry; though the high temperatures used in GC make it unsuitable for high molecular weight biopolymers or proteins (heat denatures them), frequently encountered in biochemistry, it is well suited for use in the petrochemical, environmental monitoring and remediation, and industrial chemical fields. It is also used extensively in chemistry research.

Liquid chromatography

Liquid chromatography (LC) is a separation technique in which the mobile phase is a liquid. Liquid chromatography can be carried out either in a column or a plane. Present day liquid chromatography that generally utilizes very small packing particles and a relatively high pressure is referred to as high performance liquid chromatography (HPLC).

In HPLC the sample is forced by a liquid at high pressure (the mobile phase) through a column that is packed with a stationary phase composed of irregularly or spherically shaped particles, a porous monolithic layer, or a porous membrane. HPLC is historically divided into two different sub-classes based on the polarity of the mobile and stationary phases. Methods in which the stationary phase is more polar than the mobile phase (e.g., toluene as the mobile phase, silica as the stationary phase) are termed normal phase liquid chromatography (NPLC) and the opposite (e.g., water-methanol mixture as the mobile phase and C18 = octadecylsilyl as the stationary phase) is termed reversed phase liquid chromatography (RPLC). Ironically the "normal phase" has fewer applications and RPLC is therefore used considerably more.

Specific techniques under this broad heading are listed below.

Affinity chromatography

Affinity chromatography is based on selective non-covalent interaction between an analyte and specific molecules. It is very specific, but not very robust. It is often used in biochemistry in the purification of proteins bound to tags. These fusion proteins are labeled with compounds such as His-tags, biotin or antigens, which bind to the stationary phase specifically. After purification, some of these tags are usually removed and the pure protein is obtained.

Affinity chromatography often utilizes a biomolecule's affinity for a metal (Zn, Cu, Fe, etc.). Columns are often manually prepared. Traditional affinity columns are used as a preparative step to flush out unwanted biomolecules.

Affinity chromatography can be used to:

- Purify and concentrate a substance from a mixture into a buffering solution
- Reduce the amount of a substance in a mixture
- Discern what biological compounds bind to a particular substance
- Purify and concentrate an enzyme solution.

However, HPLC techniques exist that do utilize affinity chromatogaphy properties. Immobilized Metal Affinity Chromatography (IMAC) is useful to separate aforementioned molecules based on the relative affinity for the metal (I.e. Dionex IMAC). Often these columns can be loaded with different metals to create a column with a targeted affinity.

The immobile phase is typically a gel matrix, often of agarose; a linear sugar molecule derived from algae. Usually the starting point is an undefined heterogeneous group of molecules in solution, such as a cell lysate, growth medium or blood serum. The molecule of interest will have a well known and defined property which can be exploited during the affinity purification process. The process itself can be thought of as an entrapment, with the target molecule becoming trapped on a solid or stationary phase or medium. The other molecules in solution will not become trapped as they do not possess this property. The solid medium can then be removed from the mixture, washed and the target molecule released from the entrapment in a process known as elution. Possibly the most common use of affinity chromatography is for the purification of recombinant proteins.

Supercritical fluid Chromatography

Supercritical Fluid Chromatography (SFC) is a form of normal phase chromatography that is used for the analysis and purification of low to moderate molecular weight, thermally labile molecules. It can also be used for the separation of chiral compounds. Principles are similar to those of high performance liquid chromatography (HPLC), however SFC typically utilizes carbon dioxide as the mobile phase; therefore the entire chromatographic flow path must be pressurized. Supercritical fluid chromatography is a separation technique in which the mobile phase is a fluid above and relatively close to its critical temperature and pressure. SFC finds use in industry primarily for separation of chiral molecules, and uses the same columns as standard HPLC systems. SFC is now commonly used for achiral separations and purifications in the pharmaceutical industry.

SFC utilizes carbon dioxide pumps that require that the incoming CO2 and pump heads be kept cold in order to maintain the carbon dioxide at a temperature and pressure that keeps it in a supercritical fluid state where it can be effectively metered at some specified flow rate. The chemist uses software to set mobile phase flow rate, co-solvent composition, and column temperature. In addition, SFC provides an additional control parameter - pressure - by using an automated back pressure regulator. From an operational standpoint, SFC is as simple and robust as HPLC but fraction collection is more convenient because the primary mobile phase evaporates leaving only the analyte and a small volume of polar co-solvent.

Similar to an HPLC system, SFC can use a variety of detection methods including UV/VIS, mass spectrometry, FID (unlike HPLC) and evaporative light scattering.

Techniques by separation mechanism

Ion exchange chromatography

Ion exchange chromatography (usually referred to as ion chromatography) uses an ion exchange mechanism to separate analytes based on their respective charges. It is usually performed in columns but can also be useful in planar mode. Ion exchange chromatography uses a charged stationary phase to separate charged compounds including anions, cations, amino acids, peptides, and proteins. In conventional methods the stationary phase is an ion exchange resin that carries charged functional groups that interact with oppositely charged groups of the compound to retain. Ion exchange chromatography is commonly used to purify proteins using FPLC.

Ion-exchange chromatography (or ion chromatography) is a process that allows the separation of ions and polar molecules based on their charge. It can be used for almost any kind of charged molecule including large proteins, small nucleotides and amino acids. The solution to be injected is usually called a sample, and the individually separated components are called analytes. It is often used in protein purification, water analysis, and quality control.

Size-exclusion chromatography

Size-exclusion chromatography (SEC) is a chromatographic method in which molecules in solution are separated by their size, and in some cases molecular weight. It is usually applied to large molecules or macromolecular complexes such as proteins and industrial polymers. Typically, when an aqueous solution is used to transport the sample through the column, the technique is known as gel-filtration chromatography, versus the name gel permeation chromatography, which is used when an organic solvent is used as a mobile phase. SEC is a widely used polymer characterization method because of its ability to provide good molar mass distribution (Mw) results for polymers. The advantages of this method include good separation of large molecules from the small molecules with a minimal volume of eluate, and that various solutions can be applied without interfering with the filtration process, all while preserving the biological activity of the particles to be separated. The technique is generally combined with others that further separate molecules by other characteristics, such as acidity, basicity, charge, and affinity for certain compounds. With size exclusion chromatography, there are short and well-defined separation times and narrow bands, which lead to good sensitivity. There is also no sample loss because solutes do not interact with the stationary phase. Disadvantages are, for example, that only a limited number of bands can be accommodated because the time scale of the chromatogram is short, and, in general, there has to be a 10% difference in molecular mass to have a good resolution[

Size-exclusion chromatography (SEC) is also known as gel permeation chromatography (GPC) or gel filtration chromatography and separates molecules according to their size (or more accurately according to their hydrodynamic diameter or hydrodynamic volume). Smaller molecules are able to enter the pores of the media and, therefore, molecules are trapped and removed from the flow of the mobile phase. The average residence time in the pores depends upon the effective size of the analyte molecules. However, molecules that are larger than the average pore size of the packing are excluded and thus suffer essentially no retention; such species are the first to be eluted. It is generally a low-resolution chromatography technique and thus it is often reserved for the final, "polishing" step of a purification. It is also useful for determining the tertiary structure and quaternary structure of purified proteins, especially since it can be carried out under native solution conditions.

EXPANDED BED ADSORPTION (EBA) CHROMATOGRAPHIC SEPARATION

Expanded Bed Adsorption (EBA) Chromatographic Separation captures a target protein from a crude feed stream when it passes through a chromatography column system containing adsorbent beads. With this technique the crude feedstock can be treated directly in the chromatographic column, avoiding the traditional clarification and pre-treatment steps. EBA Chromatographic Separation is highly scalable, from laboratory-based 1cm diameter columns to large production columns up to 2 meter in diameter. These columns can typically handle feed stock throughput of more than 1,000,000 liter per day with a production capacity of 1000 MT protein per year.

Special techniques

Reversed-phase chromatography

Reversed-phase chromatography is an elution procedure used in liquid chromatography in which the mobile phase is significantly more polar than the stationary phase.

Reversed-phase chromatography (also called RPC, reverse-phase chromatography, or hydrophobic chromatography) includes any chromatographic method that uses a hydrophobic stationary phase. RPC refers to liquid chromatography, rather than gas chromatography.

The term "reversed-phase" has a historical background. In the 1970s, most liquid chromatography was performed using a solid support stationary phase (also called a "column") containing unmodified silica or alumina resins. This technique is now called "normal phase chromatography". In normal phase chromatography, the stationary phase is hydrophilic and therefore has a strong affinity for hydrophilic molecules in the mobile phase. Thus, the hydrophilic molecules in the mobile phase tend to bind (or "adsorb") to the column, while the hydrophobic molecules pass through the column and are eluted first. In normal phase chromatography, hydrophilic molecules can be eluted from the column by increasing the polarity of the solution in the mobile phase.

The introduction of a technique using alkyl chains covalently bonded to the solid support created a hydrophobic stationary phase, which has a stronger affinity for hydrophobic compounds. The use of a hydrophobic stationary phase can be considered the opposite, or "reverse", of normal phase chromatography - hence the term "reversed-phase chromatography". Reversed-phase chromatography employs a polar (aqueous) mobile phase. As a result, hydrophobic molecules in the polar mobile phase tend to adsorb to the hydrophobic stationary phase, and hydrophilic molecules in the mobile phase will pass through the column and are eluted first. Hydrophobic molecules can be eluted from the column by decreasing the polarity of the mobile phase using an organic (non-polar) solvent, which reduces hydrophobic interactions. The more hydrophobic the molecule, the more strongly it will bind to the stationary phase, and the higher the concentration of organic solvent that will be required to elute the molecule.

Many of the mathematical and experimental considerations used in other chromatographic methods also apply to RPC (for example, the separation resolution is dependent on the length of the column). It can be used for the separation of a wide variety of molecules. It is not typically used for separation of proteins, because the organic solvents used in RPC can denature many proteins. For this reason, normal phase chromatography is more commonly used for separation of proteins.

Today, RPC is a frequently used analytical technique. There are a variety of stationary phases available for use in RPC, allowing great flexibility in the development of separation methods.

Two-dimensional chromatography

In some cases, the chemistry within a given column can be insufficient to separate some analytes. It is possible to direct a series of unresolved peaks onto a second column with different physico-chemical (Chemical classification) properties. Since the mechanism of retention on this new solid support is different from the first dimensional separation, it can be possible to separate compounds that are indistinguishable by one-dimensional chromatography. The sample is spotted at one corner of a square plate,developed, air-dried, then rotated by 90° and usually redeveloped in a second solvent system.

Pyrolysis gas chromatography

Pyrolysis gas chromatography mass spectrometry is a method of chemical analysis in which the sample is heated to decomposition to produce smaller molecules that are separated by gas chromatography and detected using mass spectrometry.

Pyrolysis is the thermal decomposition of materials in an inert atmosphere or a vacuum. The sample is put into direct contact with a platinum wire, or placed in a quartz sample tube, and rapidly heated to 600–1000 °C. Depending on the application even higher temperatures are used. Three different heating techniques are used in actual pyrolyzers: Isothermal furnace, inductive heating (Curie Point filament), and resistive heating using platinum filaments. Large molecules cleave at their weakest points and produce smaller, more volatile fragments. These fragments can be separated by gas chromatography. Pyrolysis GC chromatograms are typically complex because a wide range of different decomposition products is formed. The data can either be used as fingerprint to prove material identity or the GC/MS data is used to identify individual fragments to obtain structural information. To increase the volatility of polar fragments, various methylating reagents can be added to a sample before pyrolysis.

Besides the usage of dedicated pyrolyzers, pyrolysis GC of solid and liquid samples can be performed directly inside Programmable Temperature Vaporizer (PTV) injectors that provide quick heating (up to 30 °C/s) and high maximum temperatures of 600–650 °C. This is sufficient for some pyrolysis applications. The main advantage is that no dedicated instrument has to be purchased and pyrolysis can be performed as part of routine GC analysis. In this case quartz GC inlet liners have to be used. Quantitative data can be acquired, and good results of derivatization inside the PTV injector are published as well.

Fast protein liquid chromatography

Fast protein liquid chromatography (FPLC) is a term applied to several chromatography techniques which are used to purify proteins. Many of these techniques are identical to those carried out under high performance liquid chromatography, however use of FPLC techniques are typically for preparing large scale batches of a purified product.

Fast protein liquid chromatography (FPLC), is a form of liquid chromatography that is often used to analyze or purify mixtures of proteins. As in other forms of chromatography, separation is possible because the different components of a mixture have different affinities for two materials, a moving fluid (the mobile phase)mobile phase can contain imidazole which passes through a porous solid (the stationary phase). In FPLC the mobile phase is an aqueous solution, or "buffer". The buffer flow rate is controlled by a positive-displacement pump and is normally kept constant, while the composition of the buffer can be varied by drawing fluids in different proportions from two or more external reservoirs. The stationary phase is a resin composed of beads, usually of cross-linked agarose, packed into a cylindrical glass or plastic column. FPLC resins are available in a wide range of bead sizes and surface ligands depending on the application.

In the most common FPLC strategy, ion exchange, a resin is chosen so that the protein of interest will bind to the resin by a charge interaction while in buffer A (the running buffer) but become dissociated and return to solution in buffer B (the elution buffer). A mixture containing one or more proteins of interest is dissolved in 100% buffer A and pumped into the column. The proteins of interest bind to the resin while other components are carried out in the buffer. The total flow rate of the buffer is kept constant; however, the proportion of Buffer B (the "elution" buffer) is gradually increased from 0% to 100% according to a programmed change in concentration (the "gradient"). At some point during this process each of the bound proteins dissociates and appears in the effluent. The effluent passes through two detectors which measure salt concentration (by conductivity) and protein concentration (by absorption of ultraviolet light at a wavelength of 280nm). As each protein is eluted it appears in the effluent as a "peak" in protein concentration and can be collected for further use.

FPLC was developed and marketed in Sweden by Pharmacia in 1982 and was originally called fast performance liquid chromatography to contrast it with HPLC or high-performance liquid chromatography. FPLC is generally applied only to proteins; however, because of the wide choice of resins and buffers it has broad application. In contrast to HPLC the buffer pressure used is relatively low, typically less than 5 bar, but the flow rate is relatively high, typically 1-5 ml/min. FPLC can be readily scaled from analysis of milligrams of mixtures in columns with a total volume of 5ml or less to industrial production of kilograms of purified protein in columns with volumes of many liters. When used for analysis of mixtures the effluent is usually collected in fractions of 1-5 ml which can be further analyzed, e.g. by MALDI mass spectrometry. When used for protein purification there may be only two collection containers, one for the purified product and one for waste.

Countercurrent chromatography

Countercurrent chromatography (CCC) is a type of liquid-liquid chromatography, where both the stationary and mobile phases are liquids. The operating principle of CCC equipment requires a column consisting of an open tube coiled around a bobbin. The bobbin is rotated in a double-axis gyratory motion (a cardioid), which causes a variable gravity (G) field to act on the column during each rotation. This motion causes the column to see one partitioning step per revolution and components of the sample separate in the column due to their partitioning coefficient between the two immiscible liquid phases used. There are many types of CCC available today. These include HSCCC (High Speed CCC) and HPCCC (High Performance CCC). HPCCC is the latest and best performing version of the instrumentation available currently.

Countercurrent chromatography (CCC) or partition chromatography is a category of chromatography techniques; laboratory techniques for the separation of mixtures, in this case liquid-liquid mixtures.

Chromatography in general is used to separate components of a mixture based on their differing affinities for mobile and stationary phases of a column. The components can then be analyzed separately by various sorts of detectors which may or may not be integrated into an apparatus. Partition chromatography is based on differences in capacity factor, k, and distribution coefficient, Kd, of the analytes using liquid stationary and mobile liquid phase. An analyte is the substance or chemical constituent that is being analyzed.

Standard column chromatography uses a solid stationary phase and a liquid mobile phase, while gas chromatography (GC) uses a liquid stationary phase on a solid support and a gaseous mobile phase. By contrast, in liquid-liquid chromatography, both the mobile and stationary phases are liquid. By eliminating solid supports, permanent adsorption of the analyte onto the column is avoided, and a near 100% recovery of the analyte can be achieved. The instrument is also easily switched between various modes of operation simply by changing solvents. With liquid-liquid chromatography, researchers are not limited by the composition of the columns commercially available for their instrument. Nearly any pair of immiscible solutions can be used in liquid-liquid chromatography, and most instruments can be operated in standard or reverse-phase modes.

Solvent costs are also generally cheaper than for High-performance liquid chromatography (HPLC), and the cost of purchasing and disposing of solid adsorbents is completely eliminated. Another advantage is that experiments conducted in the laboratory can easily be scaled to industrial volumes. When GC or HPLC is carried out with large volumes, resolution is lost due to issues with surface-to-volume ratios and flow dynamics; this is avoided when both phases are liquid.

Techniques

CCC can be thought of as occurring in three stages: mixing, settling, and separation (although they often occur continuously). Mixing of the phases is necessary so that the interface between them has a large area, and the analyte can move between the phases according to its partition coefficient.

A partition coefficient is a ratio of the amount of analyte found in each of the solvents at equilibrium and is related to the analyte's affinity for one over the other. The mobile phase is mixing with then settling from the stationary phase throughout the column. The degree of stationary phase retention (inversely proportional to the amount of stationary phase loss or "bleed" in the course of a separation) is a crucial parameter. Higher quality instruments have greater stationary phase retention. The settling time is a property of the solvent system and the sample matrix, both of which greatly influence stationary phase retention.

Droplet Countercurrent Chromatography (DCCC)

Droplet CCC is the oldest form of CCC. It uses only gravity to move the mobile phase through the stationary phase. In descending mode, droplets of the denser mobile phase and sample are allowed to fall through a column of the lighter stationary phase using only gravity.

If a less dense mobile phase is used it will rise through the stationary phase; this is called ascending mode. The eluent from one column is transferred to another; the more columns that are used, the more theoretical plates can be achieved. The disadvantage of DCCC is that flow rates are low, and poor mixing is achieved for most binary solvent systems, which makes this technique both time-consuming and inefficient.

HIGH-PERFORMANCE COUNTERCURRENT CHROMATOGRAPHY (HPCCC)

The operating principle of CCC equipment requires a column consisting of a tube coiled around a bobbin. The bobbin is rotated in a double-axis gyratory motion (a cardioid), which causes a variable gravity (G) field to act on the column during each rotation. This motion causes the column to see one partitioning step per revolution and components of the sample separate in the column due to their partitioning coefficient between the two immiscible liquid phases used.

HPCCC works in much the same way as HSCCC but with one vital difference. A seven-year R&D process that has produced HPCCC instruments that generated 240 g, compared to the 80 g of the HSCCC machines. This increase in g-level and larger bore of the column has enabled a tenfold increase in through put, due to improved mobile phase flow rates and a much higher stationary phase retention.

Countercurrent chromatography is a preparative liquid chromatography technique, however with the advent of the higher g HPCCC instruments it is now prossible to operate instruments with sample loadings as low as a few milligrams, whereas in the past 100s of milligrams have been necessary.

Major application areas for this technique include natural products purification and also drug development.

High-Speed Countercurrent Chromatography (HSCCC)

The modern era of CCC began with the development by Dr. Yoichiro Ito of the planetary centrifuge and the many possible column geometries it can support. These devices make use of a little-known means of making non-rotating connections between the stator and the rotor of a centrifuge. (It is beyond the scope of this discussion to describe the method of accomplishing this. Any of the several books available on CCC discuss it thoroughly.)

Functionally, the high-speed CCC consists of a helical coil of inert tubing which rotates on its planetary axis and simultaneously rotates eccentrically about another solar axis. (These axes can be made to coincide, but the most common or type J CCC is discussed here.) The effect is to create zones of mixing and zones of settling which progress along the helical coil at dizzying speed. This produces a highly favorable environment for chromatography.

There are numerous potential variants upon this instrument design. The most significant of these is the toroidal CCC. This instrument does not employ planetary motion. In some respects it is very like CPC, but retains the advantage of not needing rotary seals. It also employs a capillary tube instead of the larger-diameter tubes employed in the helices of the other CCC models. This capillary passage makes the mixing of two phases very thorough, despite the lack of shaking or other mixing forces. This instrument provides rapid analytical-scale separations, which can nonetheless be scaled up to either of the larger-scale CCC instruments.

Xanthanolide purification

Centrifugal Partition Chromatography (CPC)

Centrifugal Partition Chromatography (CPC) was invented in the eighties by the Japanese company Sanki Engineering Ltd, whose president was the late Kanichi Nunogaki. CPC has been extensively developed in France starting from the late nineties. CPC uses centrifugal force to speed separation and achieves higher flow rates than DCCC (which relies on gravity).

The centrifugal partition chromatograph is constituted with a unique rotor (=column). This rotor rotates on its central axis (while HSCCC column rotates on its planetary axis and simultaneously rotates eccentrically about another solar axis). With less vibrations and noise, the CPC offers a wider rotation speed range (from 500 to 2000 rpm) than HSCCC. That allows a better decantation and retention for unstable biphasic system (e.g., aqueous aqueous systems or Butanol/water systems).

Basics of CPC: The CPC rotor is constituted by the superposition of disks engraved with small cells connected by head / tail ducts. These cells, where the chromatographic separation takes place, can be compared to lined-up separate funnels. The rotor is filled with the stationary phase, which stays inside the rotor thanks to the rotation speed, while the mobile phase is pumped through. CPC can be operated in either descending or ascending mode, where the direction is relative to the force generated by the rotor rather than gravity. According to the fast and permanent evolution of the cells design, the efficiency and flow rate with low back pressure are improved. The CPC offers now the direct scale up from the analytical apparatuses (few milliliters) to industrial apparatuses (some liters) for fast batch production.

Chiral chromatography

Chiral chromatography involves the separation of stereoisomers. In the case of enantiomers, these have no chemical or physical differences apart from being three-dimensional mirror images. Conventional chromatography or other separation processes are incapable of separating them. To enable chiral separations to take place, either the mobile phase or the stationary phase must themselves be made chiral, giving differing affinities between the analytes. Chiral chromatography HPLC columns (with a chiral stationary phase) in both normal and reversed phase are commercially available.

4

CLASSICAL METHODS: CHEMICAL ANALYSIS

Analytical chemistry is the study and application of instruments and methods for separating, identifying, and quantifying matter. Separation, recognition, and quantification can be used alone or in combination with other practice methods. Separation is the process of isolating analytes. Quantitative analysis decides the numerical quantity or concentration, while qualitative analysis defines analytes.

Chemical analysis is performed collectively by a team of experts with extensive knowledge of specialized techniques such as infrared absorption, emission spectroscopy, electrochemistry, or gas chromatography, and considerable knowledge of the capabilities of the method. Used by other team members. Therefore, an analytical chemist is not a chemist who measures the chemical composition of a substance. He now takes two more steps in the analysis process to solve the problem. One is the beginning (problem definition) and the other in the end (troubleshooting). This means that analytical chemists can measure anything but a well-defined chemical composition. For example, you might decide that measuring pH is better than analyzing the final product for process control. The metal surface is more important than composition determination.

Chemical engineering is the branch of engineering that applies the physical sciences (e.g., chemistry and physics) and/or life sciences (e.g., biology, microbiology and biochemistry) together with mathematics and economics to processes that convert raw materials or chemicals into more useful or valuable forms. In addition, modern chemical engineers are also concerned with pioneering valuable materials and related techniques – which are often essential to related fields such as nanotechnology, fuel cells and biomedical engineering. Within chemical engineering, two broad subgroups include 1) design, manufacture, and operation of plants and machinery in industrial chemical and related processes ("chemical process engineers"); and 2) development of new or adapted substances for products ranging from foods and beverages to cosmetics to cleaners to pharmaceutical ingredients, among many other products ("chemical product engineers").

Chemical reactions engineering involves managing plant processes and conditions to ensure optimal plant operation. Chemical reaction engineers construct models for reactor analysis and design using laboratory data and physical parameters, such as chemical thermodynamics, to solve problems and predict reactor performance.

Chemical engineers "develop economic ways of using materials and energy" as opposed to chemists who are more interested in the basic composition of materials and synthesizing products from such. Chemical engineers use chemistry and engineering to turn raw materials into usable products, such as medicine, petrochemicals and plastics. They are also involved in waste management and research. Both applied and research facets make extensive use of computers.

A chemical engineer may be involved in industry or university research where they are tasked in designing and performing experiments to create new and better ways of production, controlling pollution, conserving resources and making these processes safer. They may be involved in designing and constructing plants as a project engineer. In this field, the chemical engineer uses their knowledge in selecting plant equipment and the optimum method of production to minimize costs and increase profitability. After its construction, they may help in upgrading its equipment. They may also be involved in its daily operations.

CLASSIFICATION OF ANALYTICAL METHODS

Classical, wet chemical methods and modern instrumental methods make up analytical chemistry. Separations such as precipitation, extraction, and distillation are used in classical qualitative methods. Colour, odour, melting point, boiling point, solubility, radioactivity, and reactivity can all be used to identify a substance. Quantitative analysis that uses mass or volume adjustments to measure quantity is known as traditional quantitative analysis. Chromatography, electrophoresis, and field-flow fractionation are some of the instrumental methods that can be used to separate samples.

To conduct an analysis, the majority of traditional analytical methods rely on chemical reactions. Instrumental methods, on the other hand, usually rely on the calculation of the physical property of the analyte.

Classical Method

For Qualitative Analysis:

In classical qualitative analysis, one or more chemical reagents are added to the analyte. The identity of the analyte can be determined by studying chemical reactions and their components. The added reagents are selected to react selectively with one or a single class of chemical compounds to produce a distinct reaction product. In certain cases, the reaction product is a precipitate, a gas, or a colored substance.

Take, for example, copper(II), which reacts with ammonia to form a deep blue copper-ammonia complex. Similarly, dissolved lead(II) reacts with chromate-containing solutions to produce yellow lead chromate. The same method can be used to qualitatively test negative ions (anions) and positive ions (cations). A common example is a reaction between carbonates and strong acids, which results in the formation of carbon dioxide gas bubbles.

The Classical Method of Analysis for Organic or Inorganic Analyte:

The analyte is classified as organic or inorganic in most cases. As a result, qualitative analysis is classified as organic or inorganic. Inorganic compounds mainly contain elements other than carbon, whereas organic compounds primarily contain carbon compounds. Sugar (C12H22O11) is an organic compound, while table salt (NaCl) is an inorganic compound.

Chemical reactions between added chemical reagents and functional groups of organic molecules are involved l in classical organic qualitative research. As a result, the assay's result provides information about a portion of the organic molecule but typically does not provide enough information to fully classify it. To classify the entire molecule, other measures such as boiling points, melting points, and densities are combined with a functional group analysis.

The reaction between bromine in a carbon tetrachloride solution and organic compounds containing carbon-carbon double bonds is an example of a chemical reaction that can be used to classify organic functional groups. Owing to the addition of bromine through the double bonds, the distinctive red-brown colour of bromine vanishes, indicating the existence of a carbon-carbon double bond.

For Quantitative Analysis

Gravimetric and volumetric analysis are two types of traditional quantitative analysis. Both approaches rely on lengthy chemical reactions involving the analyte and additional reagents.

1. Gravimetric Analysis

A precipitate is formed when an excess of added reagent reacts with the analyte. The filtrate is measured after it has been washed, dried, and weighed. The concentration or quantity of the assayed material in the analyte is calculated using its mass.

2. Volumetric Analysis

The titrimetric analysis is another name for volumetric analysis. From a buret, the reagent (the titrant) is progressively or stepwise applied to the analyte. The secret to good titrimetric analysis is to consider the titration's equivalence point (the point at which the quantities of the two reacting species are equivalent), which is usually visible as a colour transition.

SPECTROSCOPY

Spectroscopy is a technique for determining how molecules deal with electromagnetic radiation. Spectroscopy encompasses a wide range of techniques, including atomic absorption spectroscopy, atomic emission spectroscopy, ultraviolet-visible spectroscopy, x-ray spectroscopy, fluorescence spectroscopy, infrared spectroscopy, Raman spectroscopy, dual-polarization interferometry, nuclear magnetic resonance spectroscopy, and photoemission spectroscopy. Some of the spectroscopy is discussed below.

Nuclear Magnetic Resonance

Different spectral regions of absorption correspond to different physical processes occurring within the analyte. In the presence of a magnetic field, absorption of energy in the radiofrequency region is sufficient to cause a spinning nucleus in certain atoms to shift to a different spin state. As a result, nuclear magnetic resonance spectrometry may be used to determine the number and types of various nuclei present in the groups attached to the atom containing the nucleus under investigation. It's particularly useful for analyzing organic compounds qualitatively.

Atomic Absorption Spectroscopy

Atomic absorption spectroscopy (AAS) and atomic emission spectroscopy (AES) are Spectro analytical procedures that use the absorption of optical radiation (light) by free atoms in the gaseous state to determine chemical elements quantitatively. The absorption of light by free metallic ions is the basis for atomic absorption spectroscopy.

The technique is used in analytical chemistry to determine the concentration of a certain element (the analyte) in a sample to be analyzed.

Ultraviolet-Visible Spectrophotometry

UV–Vis spectroscopy, also known as ultraviolet-visible spectrophotometry (UV–Vis or UV/Vis), is the study of absorption and reflectance spectroscopy in the ultraviolet and visible regions of the electromagnetic spectrum. This means it makes use of visible and neighboring light.

Fluorescence Spectroscopy

Fluorescence spectroscopy (also known as fluorimetry or spectrofluorometry) is an electromagnetic spectroscopy technique for analyzing fluorescence in a sample. It entails the use of a beam of light, generally ultraviolet light, to excite the electrons in certain compounds' molecules and cause them to emit light; normally, though not always, visible light.

Magnetic Resonance Spectroscopy

The spectroscopic technique of nuclear magnetic resonance spectroscopy, also known as NMR spectroscopy or magnetic resonance spectroscopy (MRS), is used to observe local magnetic fields around atomic nuclei. The Nuclear magnetic resonance signal is provided by excitation of the nuclei sample with radio waves into nuclear magnetic resonance, which is detected with sensitive radio receivers, and the sample is placed in a magnetic field. The resonance frequency of an atom in a molecule is changed by the intramolecular magnetic field surrounding it, allowing access to knowledge of the molecule's electronic structure and individual functional groups.

The spectroscopic technique of nuclear magnetic resonance spectroscopy, also known as Nuclear magnetic resonance spectroscopy or magnetic resonance spectroscopy (MRS), is used to observe local magnetic fields around atomic nuclei. The NMR signal is provided by excitation of the nuclei sample with radio waves into nuclear magnetic resonance, which is detected with sensitive radio receivers, and the sample is placed in a magnetic field. The resonance frequency of an atom in a molecule is changed by the intramolecular magnetic field surrounding it, allowing access to knowledge of the molecule's electronic structure and individual functional groups.

Electrochemical Analysis

The potential (volts) and/or current (amps) in an electrochemical cell containing the analyte are measured using electroanalytical methods. These techniques are classified based on which aspects of the cell are regulated and which are calculated. Potentiometry (measuring the difference in electrode potentials), coulometry (measuring the transferred charge over time), amperometry (measuring the cell's current over time), and voltammetry (measuring the cell's current while actively altering the cell's potential) are the four major groups.

Thermal Analysis

The interaction of a substance and heat is measured using calorimetry and thermogravimetric analysis.

Separation

Separation methods are used to make content mixtures less complicated. This field includes chromatography, electrophoresis, and field flow fractionation.

Microscopy

In analytical science, visualization of single molecules, single cells, biological tissues, and nanomaterials is an effective and appealing approach. Analytical science is now being revolutionized by the hybridization of other conventional analytical methods. Optical microscopy, electron microscopy, and scanning probe microscopy are the three different types of microscopy. Because of the rapid growth of the computer and camera industries, this sector has recently accelerated.

PRECIPITATION ANALYSIS

Cations are generally classified according to their properties into one of six groups and anions into one of three groups. The details of the classification vary slightly from source to source. Each group has a common reagent that can be used to separate cations or anions in solution. Separations must be performed in a specific order to achieve meaningful results. This is because some ions from the former group may react with reagents from the latter group. This is because cation analysis is based on ionic solubility products. When the cations reach the desired optimal concentration, they are precipitated for detection. Copper in the precipitation analysis wire displaces silver, silver is immersed in silver nitrate solution, and precipitates solid silver. Sediment analysis can be used to determine the chemical composition of a solution. The cationic group includes:

Cation forms insoluble chlorides such as lead, silver, and mercury. Formation of acid-insoluble sulfides such as cadmium, bismuth, copper, antimony, and tin. With the formation of insoluble hydroxide complexes such as iron, aluminum and chromium.

Adding ammonia chloride, ammonium hydroxide, and hydrogen sulfide gases to zinc, nickel, cobalt, and manganese will determine the element. The colour of the sediment indicates the metal. Insoluble carbonate group. (Many initial cations precipitate as carbonate, but will be detected up to this point if you follow the steps in a certain order.) Barium, calcium, and strontium precipitate at this point, but not earlier. Magnesium, lithium, sodium, potassium, and ammonium are difficult to precipitate and are usually identified by the colour of the flame. There are three groups of negative ions, and there are several ways to detect them. However, deposition methods such as those mentioned above are often used.

COLORIMETRIC ANALYSIS

Using a coloured reagent, colorimetric analysis measures the concentration of a chemical in a solution. Inorganic and organic compounds are examined in this way. In medical laboratories and industrial water treatment, this method is widely used to analyze water samples.

What are the Similarities Between Qualitative Analysis and Quantitative Analysis in Chemistry?

Chemical analysis can be categorized as either qualitative or quantitative. Chemical analyses both qualitatively and quantitatively reveal details about sample components. The quantitative and qualitative methods are often applied in conjunction with each other. The main analytical method in chemistry for quantifying and qualitatively determining the chemical composition of a sample is Qualitative and Quantitative Chemical Analysis.

Uses of Chemical Analysis:

As a result, analytical chemistry has a vital role in medicine, as it can be used to determine the amounts of ingredients such as carbohydrates, lipids, proteins, and sugars in the body. As well as determining levels of toxic wastes in the body, such as uric acid, cholesterol, drugs, and certain salts, analytical methods can help determine levels of toxic wastes in the body.

Identifying unknown substances and matching samples to known substances is a task for forensic chemists when they analyze abiotic samples found at crime scenes. Furthermore, we identify and sometimes quantify drug/regulated substances found in the field of human beings.

Titration

Titration, also known as titrimetry, is a common laboratory method of quantitative chemical analysis that is used to determine the unknown concentration of an identified analyte. Because volume measurements play a key role in titration, it is also known as volumetric analysis. A reagent, called the titrant or titrator is prepared as a standard solution. A known concentration and volume of titrant reacts with a solution of analyte or titrand to determine concentration. The word "titration" comes from the Latin word titulus, meaning inscription or title. The French word titre, also from this origin, means rank. Titration, by definition, is the determination of rank or concentration of a solution with respect to water with a pH of 7 (the pH of pure H2O under standard conditions).

Volumetric analysis originated in late 18th-century France. François-Antoine-Henri Descroizilles developed the first burette (which was similar to a graduated cylinder) in 1791. Joseph Louis Gay-Lussac developed an improved version of the burette that included a side arm, and coined the terms "pipette" and "burette" in an 1824 paper on the standardization of indigo solutions. A major breakthrough in the methodology and popularization of volumetric analysis was due to Karl Friedrich Mohr, who redesigned the burette by placing a clamp and a tip at the bottom, and wrote the first textbook on the topic, Lehrbuch der chemisch-analytischen Titrirmethode (Textbook of analytical-chemical titration methods), published in 1855.

Procedure

A typical titration begins with a beaker or Erlenmeyer flask containing a precise volume of the titrand and a small amount of indicator placed underneath a calibrated burette or chemistry pipetting syringe containing the titrant. Small volumes of the titrant are then added to the titrand and indicator until the indicator changes, reflecting arrival at the endpoint of the titration. Depending on the endpoint desired, single drops or less than a single drop of the titrant can make the difference between a permanent and temporary change in the indicator.

Preparation techniques

Typical titrations require titrant and analyte to be in a liquid (solution) form. Though solids are usually dissolved into an aqueous solution, other solvents such as glacial acetic acid or ethanol are used for special purposes (as in petrochemistry). Concentrated analytes are often diluted to improve accuracy.

Many non-acid-base titrations require a constant pH throughout the reaction. Therefore a buffer solution may be added to the titration chamber to maintain the pH.

In instances where two reactants in a sample may react with the titrant and only one is the desired analyte, a separate masking solution may be added to the reaction chamber which masks the unwanted ion.

Some redox reactions may require heating the sample solution and titrating while the solution is still hot to increase the reaction rate. For instance, the oxidation of some oxalate solutions requires heating to 60 °C (140 °F) to maintain a reasonable rate of reaction.

Titration curves

A titration curve is a curve in the plane whose x-coordinate is the volume of titrant added since the beginning of the titration, and whose y-coordinate is the concentration of the analyte at the corresponding stage of the titration (in an acid-base titration, the y-coordinate is usually the pH of the solution).

In an acid-base titration, the titration curve reflects the strength of the corresponding acid and base. For a strong acid and a strong base, the curve will be relatively smooth and very steep near the equivalence point. Because of this, a small change in titrant volume near the equivalence point results in a large pH change and many indicators would be appropriate (for instance litmus, phenolphthalein or bromothymol blue).

If one reagent is a weak acid or base and the other is a strong acid or base, the titration curve is irregular and the pH shifts less with small additions of titrant near the equivalence point. For example, the titration curve for the titration between oxalic acid (a weak acid) and sodium hydroxide (a strong base) is pictured. The equivalence point occurs between pH 8-10, indicating the solution is basic at the equivalence point and an indicator such as phenolphthalein would be appropriate. Titration curves corresponding to weak bases and strong acids are similarly behaved, with the solution being acidic at the equivalence point and indicators such as methyl orange and bromothymol blue being most appropriate.

Titrations between a weak acid and a weak base have titration curves which are highly irregular. Because of this, no definite indicator may be appropriate and a pH meter is often used to monitor the reaction.

The type of function that can be used to describe the curve is called a sigmoid function.

Types of titrations

There are many types of titrations with different procedures and goals. The most common types of qualitative titration are acid-base titrations and redox titrations.

Acid–base titration

An acid-base titration is the determination of the concentration of an acid or base by exactly neutralizing the acid or base with an acid or base of known concentration. This allows for quantitative analysis of the concentration of an unknown acid or base solution. It makes use of the neutralization reaction that occurs between acids and bases and the knowledge of how acids and bases will react if their formulas are known.

Acid–base titrations can also be used to find percent purity of chemicals.

Indicator	Color on acidic side	Range of color change	Color on basic side
Methyl Violet	Yellow	0.0–1.6	Violet
Bromophenol Blue	Yellow	3.0–4.6	Blue
Methyl Orange	Red	3.1–4.4	Yellow
Methyl Red	Red	4.4–6.3	Yellow
Litmus	Red	5.0–8.0	Blue
Bromothymol Blue	Yellow	6.0–7.6	Blue
Phenolphthalein	Colorless	8.3–10.0	Pink
Alizarin Yellow	Yellow	10.1–12.0	Red

Acid-base titrations depend on the neutralization between an acid and a base when mixed in solution. In addition to the sample, an appropriate indicator is added to the titration chamber, reflecting the pH range of the equivalence point. The acid-base indicator indicates the endpoint of the titration by changing color. The endpoint and the equivalence point are not exactly the same because the equivalence point is determined by the stoichiometry of the reaction while the endpoint is just the color change from the indicator. Thus, a careful selection of the indicator will reduce the indicator error. For example, if the equivalence point is at a pH of 8.4, then the Phenolphthalein indicator would be used instead of Alizarin Yellow because phenolphthalein would reduce the indicator error. Common indicators, their colors, and the pH range in which they change color are given in the table above. When more precise results are required, or when the reagents are a weak acid and a weak base, a pH meter or a conductance meter are used.

Alkalimetry is the specialized analytic use of acid-base titration to determine the concentration of a basic (synonymous to alkaline) substance. Acidimetry, sometimes spelled acidometry, is the same concept of specialized analytic acid-base titration, but for an acidic substance.

Before starting the titration a suitable pH indicator must be chosen. The equivalence point of the reaction, the point at which equivalent amounts of the reactants have reacted, will have a pH dependent on the relative strengths of the acid and base used. The pH of the equivalence point can be estimated using the following rules:

- A strong acid will react with a strong base to form a neutral (pH = 7) solution.
- A strong acid will react with a weak base to form an acidic (pH < 7) solution.
- A weak acid will react with a strong base to form a basic (pH > 7) solution.

When a weak acid reacts with a weak base, the equivalence point solution will be basic if the base is stronger and acidic if the acid is stronger. If both are of equal strength, then the equivalence pH will be neutral. However, weak acids are not often titrated against weak bases because the colour change shown with the indicator is often quick, and therefore very difficult for the observer to see the change of colour.

The point at which the indicator changes colour is called the end point. A suitable indicator should be chosen, preferably one that will experience a change in colour (an end point) close to the equivalence point of the reaction.

First, the burette should be rinsed with the standard solution, the pipette with the unknown solution, and the conical flask with distilled water.

Secondly, a known volume of the unknown concentration solution should be taken with the pipette and placed into the conical flask, along with a small amount of the indicator chosen.

The known solution should then be allowed out of the burette, into the conical flask. At this stage we want a rough estimate of the amount of this solution it took to neutralize the unknown solution. The solution should be let out of the burette until the indicator changes colour and the value on the burette should be recorded. This is the first (or rough) titre and should be discluded from any calculations.

At least three more titrations should be performed, this time more accurately, taking into account roughly where the end point will occur. The initial and final readings on the burette (prior to starting the titration and at the end point, respectively) should be recorded. Subtracting the initial volume from the final volume will yield the amount of titrant used to reach the end point. The end point is reached when the indicator just changes colour permanently. This is best achieved by washing a hanging drop from the tip of the burette into the flask right at the end of the titration to achieve a drop that is smaller in volume than what can usually be achieved by just dripping solution off the burette.

Acid–base titration is performed with a phenolphthalein indicator, when it is a strong acid – strong base titration, a bromthymol blue indicator in weak acid – strong base reactions, and a methyl orange indicator for strong acid – weak base reactions. If the base is off the scale, i.e. a pH of >13.5, and the acid has a pH >5.5, then an Alizarine yellow indicator may be used. On the other hand, if the acid is off the scale, i.e. a pH of <0.5, and the base has a pH <8.5, then a Thymol Blue indicator may be used.

Redox titration

Redox titration (also called oxidation-reduction titration) is a type of titration based on a redox reaction between the analyte and titrant. Redox titration may involve the use of a redox indicator and/or a potentiometer. Redox titrations are based on a reduction-oxidation reaction between an oxidizing agent and a reducing agent. A potentiometer or a redox indicator is usually used to determine the endpoint of the titration, as when one of the constituents is the oxidizing agent potassium dichromate. The color change of the solution from orange to green is not definite, therefore an indicator such as sodium diphenylamine is used. Analysis of wines for sulfur dioxide requires iodine as an oxidizing agent. In this case, starch is used as an indicator; a blue starch-iodine complex is formed in the presence of excess iodine, signalling the endpoint.

Some redox titrations do not require an indicator, due to the intense color of the constituents. For instance, in permanganometry a slight faint persisting pink color signals the endpoint of the titration because of the color of the excess oxidizing agent potassium permanganate.

An example of a redox titration is treating a solution of iodine with a reducing agent and using starch as indicator. Iodine forms an intensely blue complex with starch. Iodine (I2) can be reduced to iodide (I”) by e.g. thiosulphate (S2O32”), and when all iodine is spent the blue colour disappears. This is called an iodometric titration.

Most often of all, the reduction of iodine to iodide is the last step in a series of reactions where the initial reactions are used to convert an unknown amount of the solute (the substance you want to analyze) to an equivalent amount of iodine, which may then be titrated. Sometimes other halogens (or halogenoalkanes) than iodine are used in the intermediate reactions because they are available in better measurable standard solutions and/or react more readily with the solute. The extra steps in iodometric titration may be worth while because the equivalence point, where the blue turns a bit colourless, is more distinct than some other analytical or may be by volumetric methods.

A redox indicator (also called an oxidation-reduction indicator) is an indicator that undergoes a definite color change at a specific electrode potential.

The requirement for fast and reversible color change means that the oxidation-reduction equilibrium for an indicator redox system needs to be established very fast. Therefore only a few classes of organic redox systems can be used for indicator purposes.

There are two common type of redox indicators:

- metal-organic complexes (Ex. phenanthroline)
- true organic redox systems (Ex. Methylene blue)

Sometimes colored inorganic oxidants or reductants (Ex. Potassium manganate, Potassium dichromate) are also incorrectly called redox indicators. They can't be classified as true redox indicators because of their irreversibility.

Almost all redox indicators with true organic redox systems involve a proton as a participant in their electrochemical reaction. Therefore sometimes redox indicators are also divided into two general groups: independent or dependent on pH.

Assay

An assay is a form of biological titration used to determine the concentration of a virus or bacterium. Serial dilutions are performed on a sample in a fixed ratio (such as 1:1, 1:2, 1:4, 1:8, etc.) until the last dilution does not give a positive test for the presence of the virus. This value is known as the titer, and is most commonly determined through enzyme-linked immunosorbent assay (ELISA).

An assay is an investigative (analytic) procedure in laboratory medicine, pharmacology, environmental biology, and molecular biology for qualitatively assessing or quantitatively measuring the presence or amount or the functional activity of a target entity (the analyte), which can be a drug or biochemical substance or a cell in an organism or organic sample. The measured entity is generally called the analyte, or the measurand or the target of the assay. The assay usually aims to measure an intensive property of the analyte and express it in the relevant measurement unit (e.g. molarity, density, functional activity in enzyme international units, degree of some effect in comparison to a standard, etc.).

If the assay involves addition of exogenous reactants (the reagents), their quantities are kept fixed (or in excess) so that the quantity (and quality) of the target is the only limiting factor for the reaction/assay process, and the difference in the assay outcome is used to deduce the unknown quality or quantity of the target in question. Some assays (e.g., biochemical assays) may be similar to or have overlap with chemical analysis and titration. But generally, assays involve biological material or phenomena which tend to be intrinsically more complex either in composition or in behavior or both. Thus reading of an assay may be quite noisy and may involve greater difficulties in interpretation than an accurate chemical titration. On the other hand, older generation qualitative assays, especially bioassays, may be much more gross and less quantitative (e.g., counting death or dysfunction of an organism or cells in a population, or some descriptive change in some body part of a group of animals).

General steps of any assay

An assay (analysis) is never an isolated process and needs to be preceded by certain necessary procedures which are the preanalytic steps and must be followed by certain necessary post analytic steps. The information communication (e.g. request to perform an assay and further information processing) or specimen handling (e.g. collection transport and processing) done before and till the point of beginning of an assay are the preanalytic steps. Similarly after the actual assay is done the result may be documented, verified and transmitted/communicated in steps which are called post-analytic steps related to an assay. Like any multistep information handling and transmission systems, variation and errors in the communicated final results of an assay involves corresponding parts in every such step i.e. not only analytic variations and errors intrinsic to the assay itself but also variations and errors involved in preanalytic and post analytic steps. Since the assay itself (the analytic step) gets a lot of attention, steps that get less attention by the chain of users i.e. the preanalytic and the post analytic steps are often less stringently regulated and generally more prone to errors- e.g. preanalytic steps in medical laboratory assays may contribute to 32-75% of all lab errors.

Assays can be very diverse, but generally involve the following general steps:

1. Sample processing/manipulation in order to selectively present that target in a discernible/measurable form to a discrimination/identification/detection system. It might involve a simple centrifugal separation or washing or filtration or capture by some form of selective binding or it may even involve modifying the target e.g. epitope retrieval in immunological assays or cutting down the target into pieces e.g. in Mass Spectrometry. Generally there are multiple separate steps done before an assay and are called preanalytic processing. But some of the manipulations may be inseparable part of the assay itself and will not thus be considered pre-analytic.

2. Target specific discrimination/identification principle: to discriminate from background (noise) of similar components and specifically identify a particular target component ("analyte") in a biological material by its specific attributes. (e.g. in a PCR assay a specific oligonucleotide primer identifies the target by base pairing based on the specific nucleotide sequence unique to the target).

3. Signal (or target) AMPLIFICATION System: The presence and quantity of that analyte is converted into a detectable signal generally involving some method of signal amplification, so that it can be easily discriminated from noise and measured - e.g. in a PCR assay among a mixture of DNA sequences only the specific target is amplified into millions of copies by a DNA polymerase enzyme so that it can be discerned as a more prominent component compared to any other potential components. Sometimes the concentration of the analyte is too large and in that case the assay may involve sample dilution or some sort of signal diminution system which is a negative amplification.

4. Signal DETECTION (and interpretation) system: A system of deciphering the amplified signal into an interpretable output that can be quantitative or qualitative. It can be visual or manual very crude methods or can be vary sophisticaed electronic digital or analog detectors.

5. Signal enhancement and noise filtering: may be done at any/all of the steps above. Since the more downstream a step/process during an assay, the higher the chance of carrying over noise from the previous process and amplifying it, multiple steps in a sophisticated assay might involve various means of signal-specific sharpening/enhancement arrangements and noise reduction or filtering arrangements. These may simply be in the form of a narrow band-pass optical filer, or a blocking reagent in a binding reaction that prevents nonspecific binding or a quenching reagent in a fluorescence detection system that prevents "autofluorescence" of background objects.

Measuring the endpoint of a titration

Different methods to determine the endpoint include:

- Indicator: A substance that changes color in response to a chemical change. An acid-base indicator (e.g., phenolphthalein) changes color depending on the pH. Redox indicators are also used. A drop of indicator solution is added to the titration at the beginning; the endpoint has been reached when the color changes.
- Potentiometer: An instrument that measures the electrode potential of the solution. These are used for redox titrations; the potential of the working electrode will suddenly change as the endpoint is reached.
- pH meter: A potentiometer with an electrode whose potential depends on the amount of H+ ion present in the solution. (This is an example of an ion-selective electrode.) The pH of the solution is measured throughout the titration, more accurately than with an indicator; at the endpoint there will be a sudden change in the measured pH.

- Conductivity: A measurement of ions in a solution. Ion concentration can change significantly in a titration, which changes the conductivity. (For instance, during an acid-base titration, the H+ and OH- ions react to form neutral H2O.) As total conductance depends on all ions present in the solution and not all ions contribute equally (due to mobility and ionic strength), predicting the change in conductivity is more difficult than measuring it.
- Color change: In some reactions, the solution changes color without any added indicator. This is often seen in redox titrations when the different oxidation states of the product and reactant produce different colors.
- Precipitation: If a reaction produces a solid, a precipitate will form during the titration. A classic example is the reaction between Ag+ and Cl- to form the insoluble salt AgCl. Cloudy precipitates usually make it difficult to determine the endpoint precisely. To compensate, precipitation titrations often have to be done as "back" titrations (see below).
- Isothermal titration calorimeter: An instrument that measures the heat produced or consumed by the reaction to determine the endpoint. Used in biochemical titrations, such as the determination of how substrates bind to enzymes.
- Thermometric titrimetry: Differentiated from calorimetric titrimetry because the heat of the reaction (as indicated by temperature rise or fall) is not used to determine the amount of analyte in the sample solution. Instead, the endpoint is determined by the rate of temperature change.
- Spectroscopy: Used to measure the absorption of light by the solution during titration if the spectrum of the reactant, titrant or product is known. The concentration of the material can be determined by Beer's Law.
- Amperometry: Measures the current produced by the titration reaction as a result of the oxidation or reduction of the analyte. The endpoint is detected as a change in the current. This method is most useful when the excess titrant can be reduced, as in the titration of halides with Ag+.

Endpoint and equivalence point

Though equivalence point and endpoint are used interchangeably, they are different terms. Equivalence point is the theoretical completion of the reaction: the volume of added titrant at which the number of moles of titrant is equal to the number of moles of analyte, or some multiple thereof (as in polyprotic acids).

Endpoint is what is actually measured, a physical change in the solution as determined by an indicator or an instrument mentioned above.

There is a slight difference between the endpoint and the equivalence point of the titration. This error is referred to as an indicator error, and it is indeterminate.

Back titration

Back titration is a titration done in reverse; instead of titrating the original sample, a known excess of standard reagent is added to the solution, and the excess is titrated. A back titration is useful if the endpoint of the reverse titration is easier to identify than the endpoint of the normal titration, as with precipitation reactions. Back titrations are also useful if the reaction between the analyte and the titrant is very slow, or when the analyte is in a non-soluble solid.

Particular uses

Specific examples of titrations include:

Acid-Base Titrations

- In biodiesel: Waste vegetable oil (WVO) must be neutralized before a batch may be processed. A portion of WVO is titrated with a base to determine acidity, so the rest of the batch may be properly neutralized. This removes free fatty acids from the WVO that would normally react to make soap instead of biodiesel.
- Kjeldahl method: A measure of nitrogen content in a sample. Organic nitrogen is digested into ammonia with sulfuric acid and potassium sulfate. Finally, ammonia is back titrated with boric acid and then sodium carbonate.
- Acid value: The mass in milligrams of potassium hydroxide (KOH) required to neutralize carboxylic acid in one gram of sample. An example is the determination of free fatty acid content. These titrations are achieved at low temperatures.
- Saponification value: The mass in milligrams of KOH required to saponify carboxylic acid in one gram of sample. Saponification is used to determine average chain length of fatty acids in fat. These titrations are achieved at high temperatures.
- Ester value (or ester index): A calculated index. Ester value = Saponification value – Acid value.
- Amine value: The mass in milligrams of KOH equal to the amine content in one gram of sample.

- Hydroxyl value: The mass in milligrams of KOH required to neutralize hydroxyl groups in one gram of sample. The analyte is acetylated using acetic anhydride then titrated with KOH.

Redox titrations

- Winkler test for dissolved oxygen: Used to determine oxygen concentration in water. Oxygen in water samples is reduced using manganese(II) sulfate, which reacts with potassium iodide to produce iodine. The iodine is released in proportion to the oxygen in the sample, thus the oxygen concentration is determined with a redox titration of iodine with thiosulfate using a starch indicator.
- Vitamin C: Also known as ascorbic acid, vitamin C is a powerful reducing agent. Its concentration can easily be identified when titrated with the blue dye Dichlorophenolindophenol (DCPIP) which turns colorless when reduced by the vitamin.
- Benedict's reagent: Excess glucose in urine may indicate diabetes in the patient. Benedict's method is the conventional method to quantify glucose in urine using a prepared reagent. In this titration, glucose reduces cupric ions to cuprous ions which react with potassium thiocyanate to produce a white precipitate, indicating the endpoint.
- Bromine number: A measure of unsaturation in an analyte, expressed in milligrams of bromine absorbed by 100 grams of sample.
- Iodine number: A measure of unsaturation in an analyte, expressed in grams of iodine absorbed by 100 grams of sample.

Miscellaneous

- Karl Fischer titration: A potentiometric method to analyze trace amounts of water in a substance. A sample is dissolved in methanol, and titrated with Karl Fischer reagent. The reagent contains iodine, which reacts proportionally with water. Thus, the water content can be determined by monitoring the potential of excess iodine.

5

CHEMICAL FUELS

A fuel is any material that can be made to react with other substances so that it releases energy as thermal energy or to be used for work. The concept was originally applied solely to those materials capable of releasing chemical energy but has since also been applied to other sources of heat energy, such as nuclear energy (via nuclear fission and nuclear fusion).

The heat energy released by reactions of fuels can be converted into mechanical energy via a heat engine. Other times, the heat itself is valued for warmth, cooking, or industrial processes, as well as the illumination that accompanies combustion. Fuels are also used in the cells of organisms in a process known as cellular respiration, where organic molecules are oxidized to release usable energy. Hydrocarbons and related organic molecules are by far the most common source of fuel used by humans, but other substances, including radioactive metals, are also utilized.

The first known use of fuel was the combustion of wood or sticks by Homo erectus nearly two million years ago. Throughout most of human history only fuels derived from plants or animal fat were used by humans. Charcoal, a wood derivative, has been used since at least 6,000 BCE for melting metals. It was only supplanted by coke, derived from coal, as European forests started to become depleted around the 18th century. Charcoal briquettes are now commonly used as a fuel for barbecue cooking.

Crude oil was distilled by Persian chemists, with clear descriptions given in Arabic handbooks such as those of Muhammad ibn Zakariya Razi. He described the process of distilling crude oil/petroleum into kerosene, as well as other hydrocarbon compounds, in his Kitab al-Asrar (Book of Secrets). Kerosene was also produced during the same period from oil shale and bitumen by heating the rock to extract the oil, which was then distilled. Razi also gave the first description of a kerosene lamp using crude mineral oil, referring to it as the "naffatah".

The streets of Baghdad were paved with tar, derived from petroleum that became accessible from natural fields in the region. In the 9th century, oil fields were exploited in the area around modern Baku, Azerbaijan. These fields were described by the Arab geographer Abu al-Hasan 'Ali al-Mas'udi in the 10th century, and by Marco Polo in the 13th century, who described the output of those wells as hundreds of shiploads.

With the energy in the form of chemical energy that could be released through combustion, but the concept development of the steam engine in the United Kingdom in 1769, coal came into more common use as a power source. Coal was later used to drive ships and locomotives. By the 19th century, gas extracted from coal was being used for street lighting in London. In the 20th and 21st centuries, the primary use of coal is to generate electricity, providing 40% of the world's electrical power supply in 2005.

Fossil fuels were rapidly adopted during the Industrial Revolution, because they were more concentrated and flexible than traditional energy sources, such as water power. They have become a pivotal part of our contemporary society, with most countries in the world burning fossil fuels in order to produce power, but are falling out of favor due to the global warming and related effects that are caused from burning them.

Currently the trend has been towards renewable fuels, such as biofuels like alcohols.

Fuels are contrasted with other substances or devices storing potential energy, such as those that directly release electrical energy (such as batteries and capacitors) or mechanical energy (such as flywheels, springs, compressed air, or water in a reservoir).

This section briefly describes the main features of fuels. Energy from the Sun is converted into chemical energy by photosynthesis. But, as we know, when we burn dried plants or wood, producing energy in the form of heat and light, we are releasing the Sun's energy originally stored in that plant or in that wood through photosynthesis. We know that, in most of the world today, wood is not the main source of fuel. We generally use natural gas or oil in our homes, and we use mainly oil and coal to heat the water to produce the steam to drive the turbines for our huge power generation systems.

These fuels - coal, oil, and natural gas - are often referred to as fossil fuels. The various types of fuels (like liquid, solid and gaseous fuels) that are available depend on various factors such as costs, availability, storage, handling, pollution and landed boilers, furnaces and other combustion equipments.

The knowledge of the fuel properties helps in selecting the right fuel for the right purpose and for the efficient use of the fuel. Laboratory tests are generally used for assessing the nature and quality of fuels.

Fuel when burnt produces heat. Thus, the substances classified as fuel must necessarily contain one or several of the combustible elements: carbon, hydrogen, sulphur, etc. In the process of combustion, the chemical energy of fuel is converted into heat energy. To utilize the energy of fuel in most usable form, it is required to transform the fuel from its one state to another, i.e. from solid to liquid or gaseous state, liquid to gaseous state, or from its chemical energy to some other form of energy via single or many stages. In this way, the energy of fuels can be utilized more effectively and efficiently for various purposes.

Fuel is any material that stores potential energy in a form that can be practicably released and used as heat energy. The concept originally applied solely to those materials storing energy in the form of chemical energy that could be released through combustion, but the concept has since been also applied to other sources of heat energy such as nuclear energy (via nuclear fission or nuclear fusion), as well as releases of chemical energy through non-combustion oxidation (such as in cellular biology or in fuel cells). The heat energy released by many fuels is harnessed into mechanical energy via an engine. Other times the heat itself is valued for warmth, cooking, or industrial processes, as well as the illumination that comes with combustion. Fuels are also used in the cells of organisms in a process known as cellular respiration, where organic molecules are oxidized to release un-usable energy. Hydrocarbons are by far the most common source of fuel used by humans, but other substances, including radioactive metals, are also utilized. Fuels are contrasted with other methods of storing potential energy, such as those that directly release electrical energy (such as batteries and capacitors) or mechanical energy (such as flywheels, springs, compressed air, or water in a reservoir).

Fossil fuels are principally hydrocarbons with minor impurities. They are so named because they originate from the decayed and fossilized remains of plants and animals that lived millions of years ago.

Fossil fuels can be separated into three categories. The first is petroleum or oil. This is a mixture of light, simple hydrocarbons dominated by the fractions with 6 to 12 carbons but also containing some light hydrocarbons (e.g., methane and ethane). Fully half of the energy consumed in the United States is from petroleum used to produce fuels for automobiles, recreational vehicles, home heating, or industrial production.

The principal use of petroleum is the production of gasoline. Over 40% all of all production ends up consumed in automobiles and such. Smaller fractions are turned into fuel oil (27%), jet fuel (7.4%), and other miscellaneous fuels, while the small fraction (about 10%) is used for the synthesis of the thousands of petrochemicals used in our daily lives. Indeed, many food compounds and pharmaceuticals owe their synthesis to a petrochemical precursor.

The second most prominent and naturally most abundant fossil fuel is coal. Coal also originates from decayed vegetative material buried eons ago, but the process is slightly different, being less oxidizing. The resulting material still has some of the original lignin-like structure exhibiting many fused rings and a large fraction of aromatic compounds. Consequently, coal is more of a polymeric substance than petroleum and is found as a solid not a liquid. The carbon to hydrogen ratio in coal is close to 1:1 (depending upon the type of coal), whereas the carbon to hydrogen ration in petroleum is closer to the 1:2 value expected for a hydrocarbon chain.

Minable coal is defined as 50% of the coal in a seam of at least 12 in thickness. The proven reserves of minable coal are sufficient to supply the industrial needs of modern society for the next four to five hundred years. Unfortunately, as a fuel source, coal has many disadvantages. It is a very dirty fuel that produces a large amount of unburned hydrocarbon, particulate, and—most damaging of all—significant quantities of sulfur dioxide byproducts. Indeed, it is the coal burning power plants of the eastern United States that are responsible for much of the acid rain and environmental damage observed in upstate New York and eastern Canada. The other significant disadvantage of coal is that it is not liquid, making it awkward to transport and store and limiting its use in applications like automobiles. A great deal of research has been done on the liquefaction of coal but with little economically viable success.

The third major fossil fuel is natural gas . This is a generic term for the light hydrocarbon fractions found associated with most oil deposits. Natural gas is mostly methane with small quantities of ethane and other gases mixed in. It is hydrogenrich, since methane has a carbon to hydrogen ratio of 1:4. It is also an excellent fuel, burning with a high heat output and little in the way of unwanted pollution. It does produce carbon dioxide , which is a greenhouse gas, but all organic compounds also generate carbon dioxide on combustion. Natural gas is also easy to transport through pressurized pipelines.

All of this would appear to make natural gas the perfect fuel. However, it is not without its drawbacks. This includes the presence of hydrogen sulfide in some gas fields, leading to the term "sour gas." Hydrogen sulfide is the smell of rotten eggs, but if smell were the only problem, this would be of little concern. However, hydrogen sulfide is extremely corrosive to the pipes used to transport natural gas and is a very toxic compound being lethal at levels around 1,500 ppm.

In addition, natural gas is potentially explosive as the gas must be maintained under pressure, and any hydrocarbon, in a gaseous state, can explode. This is in contrast to the use of gasoline, which is a much safer fuel. Nevertheless, both the gaseous and liquid forms of hydrocarbons are much more volatile and represent a hazard compared to coal.

It has been suggested that because all of the oxygen in the atmosphere came from the splitting of carbon dioxide via photosynthesis, the total oxygen content may lead to an estimate of the total carbon reserves. However, it is not the total abundance of fuels that is critical—it is the accessibility from both an engineering and economics standpoint that makes near-term global fuel shortages appear probable.

In addition there are serious flaws in the gross estimate of fuels as a significant amount of the world's carbon reserves are tied up in calcium carbonate rock formations. Loosely speaking, calcium carbonate is limestone and there is abundance of limestone present throughout the world. Accordingly, most industry analysts place available fossil fuel reserves at much lower levels. It is estimated by a wide variety of sources that we will reach maximum oil production in the next twenty years. After that, production will decline worldwide and we will be forced to wean ourselves from an oil-based society. The estimates for coal provide a slightly better prognosis giving a window of about 500 years for consumption of all known reserves.

Natural gas also has one significant advantage over the other fossil fuels: it is "renewable." Natural gas is found as a side product of any decaying material. Methanogenic bacteria—literally, methane-making bacteria—exist in the garbage dumps and waste disposal sites of the industrial world, busily producing methane from garbage. Enough that the Fresh Kills garbage dump on Staten Island, New York is capable of heating 16,000 homes.

Various companies have been exploring the use of hydrogen as a fuel. When used in simple combustion, hydrogen has some of the problems associated with natural gas. It must be stored under pressure and is extremely explosive upon ignition in air (e.g. the Hindenburg disaster, in which a hydrogen-filled airship explosively burned). The type of explosion—the shape of the detonation—also makes hydrogen unsuitable as an alternative fuel for the conventional automobile. The sharpness of the explosion would quickly rattle pistons to pieces.

However, hydrogen does not need to be burned directly with oxygen to provide energy. Fuel cells combine hydrogen and oxygen at electrodes to produce electricity , which can then be used to run an electric motor or a spacecraft. NASA has been employing hydrogen/oxygen fuel cells for years to provide the electricity for both manned and unmanned spacecraft. In addition, fuel cells have the added bonus of providing crew members with fresh drinking water , as the only product of the reaction is pure water. Using fuel cells, hydrogen could potentially be used for conventional automobiles. Questions about storing hydrogen and long-term viability of the fuel cells need to be addressed, but the future of this technology looks promising. At present, however, our economically viable supply of hydrogen is obtained from fossil fuels because hydrogen is released from hydrocarbons during the refining process.

Both the use of sunlight and solar panels to create sufficient electricity to electrolyze water and bacteria capable of splitting water to generate hydrogen and oxygen may make hydrogen the chemical fuel of the future. In addition, research is underway into the use of methanol as a potential partner for a fuel cell, eliminating the need for hydrogen and greatly reducing the difficulties with storage and filling the tank.

The United States leads the world in energy consumption—over 90 quadrillion Btu are used annually in the United States alone (year 2001 estimates).

TYPE OF FUELS

The fuel can be classified into three type's mainly liquid, solid, and gaseous on the bases of their physical state.

Liquid Fuels

Liquid fuels like furnace oil and LSHS (low sulphur heavy stock) are predominantly used in industrial applications. The various properties of liquid fuels are given below:

Density

Density is defined as the ratio of the mass of the fuel to the volume of the fuel at a reference temperature of 15°C. Density is measured by an instrument called a hydrometer. The knowledge of density is useful for quantitative calculations and assessing ignition qualities. The unit of density is kg/m3.

Specific Gravity

This is defined as the ratio of the weight of a given volume of oil to the weight of the same volume of water at a given temperature. The density of fuel, relative to water, is called specific gravity. The specific gravity of water is defined as one. Since specific gravity is a ratio, it has no units. The measurement of specific gravity is generally made by a hydrometer. Specific gravity is used in calculations involving weights and volumes.

Viscosity

The viscosity of a fluid is a measure of its internal resistance to flow. Viscosity depends on the temperature and decreases as the temperature increases. Any numerical value for viscosity has no meaning unless the temperature is also specified. Viscosity is measured in Stokes / Centistokes. Sometimes viscosity is also quoted in Engler, Saybolt or Redwood. Each type of oil has its own temperature - viscosity relationship. The measurement of viscosity is made with an instrument called a Viscometer. Viscosity is the most important characteristic in the storage and use of fuel oil. It influences the degree of pre-heating required for handling, storage and satisfactory atomization. If the oil is too viscous, it may become difficult to pump, hard to light the burner, and difficult to handle. Poor atomization may result in the formation of carbon deposits on the burner tips or on the walls. Therefore pre-heating is necessary for proper atomization.

Flash Point

The flash point of a fuel is the lowest temperature at which the fuel can be heated so that the vapour gives off flashes momentarily when an open flame is passed over it. The flash point for furnace oil is 66 °C.

Pour Point

The pour point of a fuel is the lowest temperature at which it will pour or flow when cooled under prescribed conditions. It is a very rough indication of the lowest temperature at which fuel oil is ready to be pumped.

Specific Heat

Specific heat is the amount of calories needed to raise the temperature of 1 kg of oil by 10C. The unit of specific heat is kcal/kg0C. It varies from 0.22 to 0.28 depending on the oil specific gravity. The specific heat determines how much steam or electrical energy it takes to heat the oil to a desired temperature. Light oils have a low specific heat, whereas heavier oils have a higher specific heat.

Calorific Value

The calorific value is the measurement of heat or energy produced, and is measured either as gross calorific value or net calorific value. The difference is determined by the latent heat of condensation of the water vapour produced during the combustion process. Gross calorific value (GCV) assumes all vapour produced during the combustion process is fully condensed. Net calorific value (NCV) assumes the water leaves with the combustion products without fully being condensed. Fuels should be compared based on the net calorific value. The calorific value of coal varies considerably depending on the ash, moisture content and the type of coal while calorific value of fuel oils is much more consistent. The typical GCVs of some of the commonly used liquid fuels are given below:

Table 1. Gross calorific values for different fuel oils (adapted from Thermax India Ltd.)

Fuel Oil	Gross Calorific Value (kCal/kg)
Kerosene	11,100
Diesel Oil	-10,800
L.D.O	- 10,700
Furnace Oil	-10,500
LSHS	-10,600

Sulphur

The amount of sulphur in the fuel oil depends mainly on the source of the crude oil and to a lesser extent on the refining process. The normal sulphur content for the residual fuel oil (furnace oil) is in the order of 2 - 4 %. Typical figures for different fuel oils are shown in Table 2.

Table 2. Percentages of sulphur for different fuel oils (adapted from Thermax India Ltd.)

Fuel oil	Percentage of Sulphur
Kerosene	0.05 - 0.2
Diesel Oil	0.05 - 0.25
L.D.O	0.5 - 1.8
Furnace Oil	2.0 - 4.0
LSHS < 0.5	

The main disadvantage of sulphur is the risk of corrosion by sulphuric acid formed during and after combustion, and condensation in cool parts of the chimney or stack, air pre-heater and economizer.

ASH CONTENT

The ash value is related to the inorganic material or salts in the fuel oil. The ash levels in distillate fuels are negligible. Residual fuels have higher ash levels. These salts may be compounds of sodium, vanadium, calcium, magnesium, silicon, iron, aluminium, nickel, etc. Typically, the ash value is in the range of 0.03 - 0.07 %. Excessive ash in liquid fuels can cause fouling deposits in the combustion equipment. Ash has an erosive effect on the burner tips, causes damage to the refractories at high temperatures and gives rise to high temperature corrosion and fouling of equipments.

Carbon Residue

Carbon residue indicates the tendency of oil to deposit a carbonaceous solid residue on a hot surface, such as a burner or injection nozzle, when its vaporizable constituents evaporate. Residual oil contains carbon residue of 1 percent or more.

Water Content

The water content of furnace oil when it is supplied is normally very low because the product at refinery site is handled hot. An upper limit of 1% is specified as a standard. Water may be present in free or emulsified form and can cause damage to the inside surfaces of the furnace during combustion especially if it contains dissolved salts. It can also cause spluttering of the flame at the burner tip, possibly extinguishing the flame, reducing the flame temperature or lengthening the flame.

Storage of Fuel Oil

It can be potentially hazardous to store furnace oil in barrels. A better practice is to store it in cylindrical tanks, either above or below the ground. Furnace oil that is delivered may contain dust, water and other contaminants. The sizing of the storage tank facility is very important. A recommended storage size estimate is to provide for at least 10 days of normal consumption. Industrial heating fuel storage tanks are generally vertical mild steel tanks mounted above the ground. It is prudent for safety and environmental reasons to build bund walls around tanks to contain accidental spillages. As a certain amount of settlement of solids and sludge will occur in tanks over time, tanks should be cleaned at regular intervals: annually for heavy fuels and every two years for light fuels. Care should be taken when oil is decanted from the tanker to the storage tank. All leaks from joints, flanges and pipelines must be attended to at the earliest. Fuel oil should be free from possible contaminants such as dirt, sludge and water before it is fed to the combustion system.

SOLID FUEL (COAL)

Coal Classification

Coal is classified into three major types; anthracite, bituminous, and lignite. However, there is no clear demarcation between them. Coal is further classified as semi-anthracite, semi-bituminous, and sub-bituminous. Anthracite is the oldest coal from a geological perspective. It is a hard coal composed mainly of carbon with little volatile content and practically no moisture. Lignite is the youngest coal from a geological perspective. It is a soft coal composed mainly of volatile matter and moisture content with low fixed carbon. Fixed carbon refers to carbon in its free state, not combined with other elements. Volatile matter refers to those combustible constituents of coal that vaporize when coal is heated. The common coals used in for example Indian industry are bituminous and sub-bituminous coal. The chemical composition of coal has a strong influence on its combustibility. The properties of coal are broadly classified as physical properties and chemical properties.

Physical And Chemical Properties of Coal

Physical properties of coal include the heating value, moisture content, volatile matter and ash. The chemical properties of coal refer to the various elemental chemical constituents such as carbon, hydrogen, oxygen, and sulphur.

Analysis of CoaL

There are two methods to analyze coal: ultimate analysis and proximate analysis. The ultimate analysis determines all coal component elements, solid or gaseous and the proximate analysis determines only the fixed carbon, volatile matter, moisture and ash percentages. The ultimate analysis is determined in a properly equipped laboratory by a skilled chemist, while proximate analysis can be determined with a simple apparatus. (It may be noted that proximate has no connection with the word "approximate").

i. Measurement of moisture

The determination of moisture content is carried out by placing a sample of powdered raw coal of size 200- micron size in an uncovered crucible, which is placed in the oven kept at 108 +2 °C along with the lid. Then the sample is cooled to room temperature and weighed again. The loss in weight represents moisture.

ii. Measurement of volatile matter

A fresh sample of crushed coal is weighed, placed in a covered crucible, and heated in a furnace at 900 + 15 oC. The sample is cooled and weighed. Loss of weight represents moisture and volatile matter. The remainder is coke (fixed carbon and ash).

iii. Measurement of carbon and ash

The cover from the crucible used in the last test is removed and the crucible is heated over the Bunsen burner until all the carbon is burned. The residue is weighed, which is the incombustible ash. The difference in weight from the previous weighing is the fixed carbon. In actual practice Fixed Carbon or FC derived by subtracting from 100 the value of moisture, volatile matter and ash.

iv. Proximate analysis

The proximate analysis indicates the percentage by weight of fixed carbon, volatiles, ash, and moisture content in coal. The amounts of fixed carbon and volatile combustible matter directly contribute to the heating value of coal. Fixed carbon acts as a main heat generator during burning. High volatile matter content indicates easy ignition of fuel. The ash content is important in the design of the furnace grate, combustion volume, pollution control equipment and ash handling systems of a furnace.

v. Fixed carbon

Fixed carbon is the solid fuel left in the furnace after volatile matter is distilled off. It consists mostly of carbon but also contains some hydrogen, oxygen, sulphur and nitrogen not driven off with the gases. Fixed carbon gives a rough estimate of the heating value of coal.

vi. Volatile matter

Volatile matters are the methane, hydrocarbons, hydrogen and carbon monoxide, and incombustible gases like carbon dioxide and nitrogen found in coal. Thus the volatile matter is an index of the gaseous fuels present. A typical range of volatile matter is 20 to 35%. Volatile matter:

Proportionately increases flame length, and helps in easier ignition of coal

Sets minimum limit on the furnace height and volume

Influences secondary air requirement and distribution aspects

Influences secondary oil support

vii. Ash content

Ash is an impurity that will not burn. Typical range of ash is 5% to 40%. Ash-

Reduces handling and burning capacity.

Increases handling costs.

Affects combustion efficiency and boiler efficiency.

Causes clinkering and slagging.

viii. Moisture content

Moisture in coal must be transported, handled and stored. Since it replaces combustible matter, it decreases the heat content per kg of coal. Typical range is 0.5 to 10%. Moisture-

Increases heat loss, due to evaporation and superheating of vapour

Helps to a certain extent with binding fines

Aids radiation heat transfer

ix. Sulphur content

Typical range is 0.5 to 0.8% normally. Sulphur-

Affects clinkering and slagging tendencies

Corrodes chimney and other equipment such as air heaters and economizers

Limits exit flue gas temperature

x. The ultimate analysis indicates the various elemental chemical constituents such as carbon, hydrogen, oxygen, sulphur, etc. It is useful in determining the quantity of air required for combustion and the volume and composition of the combustion gases. This information is required for the calculation of flame temperature and the flue duct design etc.

STORAGE, HANDLING AND PREPARATION OF COAL

Uncertainty in the availability and transportation of fuel necessitates storage and subsequent handling. Storing coal has its own disadvantages like build-up of inventory, space constraints, deterioration in quality and potential fire hazards. Other minor losses associated with the storage of coal include oxidation, wind and carpet loss. A 1% oxidation of coal has the same effect as 1% ash in coal. Wind losses may account for nearly 0.5 – 1.0 % of the total loss. The main goal of good coal storage is to minimize carpet loss and the loss due to spontaneous combustion. Formation of a soft carpet, comprising of coal dust and soil, causes carpet loss. On the other hand, if the temperature gradually rises in a coal heap, then oxidation may lead to spontaneous combustion of coal stored. Carpet losses can be reduced by:

- Preparing a hard solid surface for coal to be stored
- Preparing standard storage bays of concrete and brick

In industry, coal handling methods range from manual and conveyor systems. It would be advisable to minimize the handling of coal so that further generation of fines and segregation effects are reduced. The preparation of coal prior to feeding into the boiler is an important step for achieving good combustion. Large and irregular lumps of coal may cause the following problems:

Poor combustion conditions and inadequate furnace temperature.

Higher excess air resulting in higher stack loss.

Increase of unburnts in the ash.

Low thermal efficiency.

Gaseous Fuel

Gas fuels are the most convenient because they require the least amount of handling and are used in the simplest and most maintenance-free burner systems. Gas is delivered "on tap" via a distribution network and so is suited for areas with a high population or industrial density. However, large individual consumers do have gasholders and some produce their own gas.

Types Of Gaseous Fuel

The following is a list of the types of gaseous fuel:

Fuels naturally found in nature:

Natural gas

Methane from coal mines

Fuel gases made from solid fuel

Gases derived from coal

Gases derived from waste and biomass

From other industrial processes (blast furnace gas)

Gases made from petroleum

Liquefied Petroleum gas (LPG)

Refinery gases

Gases from oil gasification

Gases from some fermentation process

Gaseous fuels in common use are liquefied petroleum gases (LPG), Natural gas, producer gas, blast furnace gas, coke oven gas etc. The calorific value of gaseous fuel is expressed in Kilocalories per normal cubic meter (kCal/Nm3) i.e. at normal temperature (20 0C) and pressure (760 mm Hg).

Properties Of Gaseous Fuels

Since most gas combustion appliances cannot utilize the heat content of the water vapour, gross calorific value is of little interest. Fuel should be compared based on the net calorific value. This is especially true for natural gas, since increased hydrogen content results in high water formation during combustion.

LPG

LPG is a predominant mixture of propane and butane with a small percentage of unsaturated (Propylene and Butylene) and some lighter C2 as well as heavier C5 fractions. Also propane (C_3H_8), Propylene (C_3H_6), normal and iso-butane (C_4H_{10}) and Butylene (C_4H_8) are included in the range of LPG. LPG may be defined as those hydrocarbons, which are gaseous at normal atmospheric pressure, but may be condensed to the liquid state at normal temperature, by the application of moderate pressures. Although they are normally used as gases, they are stored and transported as liquids under pressure for convenience and ease of handling. Liquid LPG evaporates to produce about 250 times volume of gas.

LPG vapour is denser than air: butane is about twice as heavy as air and propane about one and a half times as heavy as air. Conseque ntly, the vapour may flow along the ground and into drains sinking to the lowest level of the surroundings and be ignited at a considerable distance from the source of leakage. In still air vapour will disperse slowly. Escape of even small quantities of the liquefied gas can give rise to large volumes of vapour / air mixture and thus cause considerable hazard. To aid in the detection of atmospheric leaks, all LPG's are required to be odorized. There should be adequate ground level ventilation where LPG is stored. For this very reason LPG cylinders should not be stored in cellars or basements, which have no ventilation at ground level.

6

INSTRUMENTAL METHODS OF ANALYSIS

Instrumental methods of analysis rely on machines. There are several different types of instrumental analysis. Some are suitable for detecting and identifying elements, while others are better suited to compounds.

The instrumental methods of chemical analysis are divided into categories according to the property of the analyte that is to be measured. Many of the methods can be used for both qualitative and quantitative analysis. The major categories of instrumental methods are the spectral, electroanalytical, and separatory.

In the flame emission spectroscope, the coloured light from a vaporised sample can be split to produce an emission spectrum. The different lines in an emission spectrum look like a coloured barcode. Each metal ion produces a unique emission spectrum. The metal present in a sample is identified by comparing its spectrum with reference spectra. These are emission spectra from known metal ions. If two spectra match, they must be from the same metal ion.

The technology of using instruments to measure and control the physical and chemical properties of material is called instrumentation. Investigations in food science and technology, whether in universities, governmental agencies, or the food industry, often require determination of food composition. The dairy scientists and technologists often determine the chemical composition as part of research on food product development or quality assurance activities. The chemical composition of foods is often determined to establish the acceptability or nutritive value of food product. The field of milk and milk product analysis involves a considerable amount of time spent learning principles, methods and instrument operations and perfecting various techniques. Also the relationship between the physical properties of milk and its chemical composition has been exploited in the dairy industry to develop various instrumental methods for the determination of chemical quality or composition of milk. A variety of methods are available to assay particular components of milk and milk products. Speed, precision and accuracy are often the key factors that determine the choice of method. Process control samples are usually analyzed by rapid methods, whereas for legal requirements generally requires the use of official reference methods.

SPECTRAL METHODS

Spectral methods measure the electromagnetic radiation that is absorbed, scattered, or emitted by the analyte. Because the types of radiation that can be monitored are multitudinous and the manner in which the radiation is measured can significantly vary from one method to another, the spectral methods constitute the largest category of instrumental methods.

Spectrochemical analysis, methods of chemical analysis that depend upon the measurement of the wavelength and the intensity of electromagnetic radiation. Its major use is in the determination of the arrangement of atoms and electrons in molecules of chemical compounds on the basis of the amounts of energy absorbed during changes in the structure or motion of the molecules. In its restricted and more common usage two methods usually are implied: (1) ultraviolet (nonvisible) and visible emission spectroscopy and (2) ultraviolet, visible, and infrared absorption spectrophotometry.

In emission spectroscopy, atoms are excited to energy levels higher than their lowest normal levels (ground states) by means of electrical discharges (arcs, sparks) or flames. Identification of the elemental composition of an unknown substance is based on the fact that when the excited atoms return to lower energy states, they emit light of characteristic frequencies. These characteristic frequencies are separated into an ordered sequence (spectrum) by diffraction or refraction (deflection of the path of the light by a grating or a prism) for observation in a spectroscope (visual), spectrograph (photographic), or spectrometer (photoelectric). The process consists of four interdependent steps: (1) vaporization of the sample, (2) electronic excitation of its atoms or ions, (3) dispersion of the emitted or absorbed radiation into its component frequencies, and (4) measurement of the intensity of the radiation, usually at wavelengths at which the intensity is greatest.

Ordinarily, emission spectrochemical analysis is applied to the qualitative and quantitative determination of metallic elements, but it is not restricted to them. The method is among the most sensitive of all analytical methods: a few milligrams of a solid sample usually suffice for the detection of metallic elements present to the extent of a few parts per million or less. In addition, the method is capable of detecting several atomic species simultaneously, thus obviating chemical separations.

Quantitative analysis by emission spectroscopy depends upon the fact that the quantity of light (i.e., the intensity) emitted at a given wavelength is proportional to the number of atoms vaporized and excited. The quantity of a given element is usually determined by a comparative method—that is, the intensity of the radiation emitted at a selected wavelength by the sample is compared with the intensity of the radiation emitted by a standard of known composition. Other spectrochemical methods useful in elemental analysis are atomic absorption spectrometry and atomic fluorescence spectrometry. Both methods resemble the flame method of emission spectroscopy (i.e., a method that uses flame as the energy source to excite atoms) in that a solution of the sample is usually vaporized into a flame of hydrogen or acetylene in air or oxygen. In addition, light of the same wavelength as that emitted by the desired element is passed through the flame. A certain fraction of the light is absorbed by atoms that are in their ground electronic state. The quantity of radiation absorbed is proportional to the concentration of atoms in the flame in their ground state and, because thermal equilibrium exists, to the total concentration of that atomic species.

Atomic fluorescence spectrometry makes use of the same basic instrumental components as atomic absorption spectrometry; however, it measures the intensity of the light emitted by atoms that have been excited from their ground state by the absorption of light of shorter wavelength than that emitted. The atomic absorption method is particularly well adapted to the determination of the alkali and alkaline earth metals.

Absorptiometry

In the most often used spectral method, the electromagnetic radiation that is provided by the instrument is absorbed by the analyte, and the amount of the absorption is measured. Absorption occurs when a quantum of electromagnetic radiation, known as a photon, strikes a molecule and raises it to some excited (high-energy) state. The intensity (i.e., the energy, in the form of electromagnetic radiation, transferred across a unit area per unit time) of the incident radiation decreases as it passes through the sample. The techniques that measure absorption in order to perform an assay are absorptiometry or absorption spectrophotometry.

Normally absorptiometry is subdivided into categories depending on the energy or wavelength region of the incident radiation. In order of increasingly energetic radiation, the types of absorptiometry are radiowave absorptiometry (called nuclear magnetic resonance spectrometry), microwave absorptiometry (including electron spin resonance spectrometry), thermal absorptiometry (thermal analysis), infrared absorptiometry, ultraviolet-visible absorptiometry, and X-ray absorptiometry. The instruments that provide and measure the radiation vary from one spectral region to another, but their operating principles are the same. Each instrument consists of at least three essential components: (1) a source of electromagnetic radiation in the proper energy region, (2) a cell that is transparent to the radiation and that can contain the sample, and (3) a detector that can accurately measure the intensity of the radiation after it has passed through the cell, and the sample.

Essentially, the amount of absorbed radiation increases with the concentration of the analyte and with the distance through the analyte that the radiation must travel (the cell path length). As radiation is absorbed in the sample, the intensity of the radiative beam decreases. By measuring the decreased intensity through a fixed-path-length cell containing the sample, it is possible to determine the concentration of the sample. Because different substances absorb at different wavelengths (or energies), the instruments must be capable of controlling the wavelength of the incident electromagnetic radiation. In most instruments, this is accomplished with a monochromator. In other instruments, it is done by use of radiative filters or by use of sources that emit radiation within a narrow wavelength band.

Because the wavelength at which substances absorb radiation depends on their chemical makeup, absorptiometry can also be used for qualitative analysis. The analyte is placed in the cell, and the wavelength of the incident radiation is scanned throughout a spectral region while the absorption is measured. The resulting plot of radiative intensity or absorption as a function of wavelength or energy of the incident radiation is a spectrum. The wavelengths at which peaks are observed are used to identify components of the analyte.

Microwave Absorptiometry

In a manner that is similar to that described for nuclear magnetic resonance spectrometry, electron spin resonance spectrometry is used to study spinning electrons. The absorbed radiation falls in the microwave spectral region and induces transitions in the spin states of the electrons. An externally applied magnetic field is required. The technique is effective for studying structures and reactions of materials that contain unpaired electrons.

Absorbed microwave radiation can cause changes in rotational energy levels within molecules, making it useful for other purposes. The rotational energy levels within a molecule correspond to the different possible ways in which a portion of a molecule can revolve around the chemical bond that binds it to the remainder of the molecule. Because the permitted rotational levels depend on the natures of the bonded atoms (e.g., their masses), microwave radiation can be used for qualitative analysis of some organic molecules.

Thermal Analysis

During thermal analysis heat is added to an analyte while some property of the analyte is measured. Often the temperature of the sample is monitored during the addition of heat. The manner in which the temperature changes is compared to the way in which the temperature of a completely inert material changes while being exposed to the same heating program. The results are employed for qualitative and quantitative analysis and for determining decomposition mechanisms of the analyte. For example, compounds that contain water exhibit a constant temperature region as the water is stripped from the compound even though heat is continuously added. If the manner in which a compound responds to a heating program is known, the technique can be used for quantitative analysis by measuring the time necessary for a particular change within the analyte to occur.

Infrared Spectrophotometry

Absorbed infrared radiation causes rotational changes in molecules, as described for microwave absorption above, and also causes vibrational changes. The vibrational energy levels within a molecule correspond to the ways in which the individual atoms or groups of atoms vibrate relative to the remainder of the molecule. Because vibrational energy levels are dependent on the types of atoms and functional groups, infrared absorption spectrophotometry is primarily used for organic qualitative analysis. It can be used for quantitative analysis, however, by monitoring the amount of absorbed radiation at a given energy corresponding to one of the peaks in the spectrum of the molecule.

Ultraviolet-visible Spectrophotometry

Absorption in the ultraviolet-visible region of the spectrum causes electrons in the outermost occupied orbital of an atom or molecule to be moved to a higher (i.e., farther from the nucleus) unoccupied orbital. Ultraviolet-visible absorptiometry is principally used for quantitative analysis of atoms or molecules. It is a useful method in this respect because the height of the absorption peaks in the ultraviolet-visible region of the spectra of many organic and inorganic compounds is large in comparison to the peak heights observed in other spectral regions. Small analyte concentrations can be more easily measured when the peaks are high. If the analyte consists of discrete atoms (which exist only in the gaseous state), the method is termed atomic absorption spectrophotometry.

Some ions and molecules do not absorb strongly in the ultraviolet-visible spectral region. Methods have been developed to apply ultraviolet-visible absorptiometry to those substances. Normally a chemical reagent is added that reacts with the analyte to form a reaction product that strongly absorbs. The absorption of the product of the chemical reaction is measured and related to the concentration of the nonabsorbing analyte. When a nonabsorbing metallic ion is assayed, the added reagent generally is a complexing agent. For example, 1,10-phenanthroline is added to solutions that are assayed for iron(II). The complex that forms between the iron and the reagent is red and is suitable for determining even very small amounts of iron. When a chemical reagent is used in a spectrophotometric assay, the procedure is called a spectrochemical analysis.

Spectrophotometric titrations are another example of spectrochemical analyses. The titrant (reagent) is placed in a buret and is added stepwise to the assayed substance. After each addition, the absorption of the solution in the reaction vessel is measured. A titration curve is prepared by plotting the amount of absorption as a function of the volume of added reagent. The shape of the titration curve depends on the absorbances of the titrant, analyte, and reaction product; from the shape of the curve, it is possible to determine the end point. The end-point volume is used with the concentration of the reagent and the initial volume of the sample solution to calculate the concentration of the analyte.

The detectors that are used in ultraviolet-visible spectrophotometry measure photons. If these photon detectors are replaced by a detector that measures pressure waves, the technique is known as photoacoustic, or optoacoustic, spectrometry. Photoacoustic spectrometers typically employ microphones or piezoelectric transducers as detectors. Pressure waves result when the analyte expands and contracts as it absorbs chopped electromagnetic radiation.

X-ray Absorption

Absorbed X rays cause excitation of electrons from inner orbitals (those near the nucleus) to unoccupied outer orbitals. In some cases, the energy of the incident X ray is sufficient to ionize the analyte by completely removing the electron from the atom or molecule. The energy required to excite the electron from an inner orbital is greater than that which is available in the ultraviolet-visible region. Because the inner shell electrons that are excited during X-ray absorption are associated with atoms in molecules rather than with the molecule as a whole, the information that is provided from a study of X-ray absorption spectra relates to the atoms within a molecule rather than to the entire molecule. X-ray absorption is used for qualitative analysis by comparing the spectrum of the analyte to spectra of known substances. Quantitative analysis also is performed in a manner similar to that used in other spectral regions. X-ray absorption spectra differ in shape from those observed in other regions, but the same measurement principles are applied during the assays.

Scattered Radiation

Radiative scattering is utilized in the second major spectral method of analysis. In this technique some radiation that passes through a sample strikes particles of the analyte and is scattered in a different direction. A detector is used to measure either the intensity of the scattered radiation or the decreased intensity of the incident radiation. Depending on the scattering mechanism, the method can be employed for either qualitative or quantitative analysis. If the intensity of the scattered radiation is measured, quantitative analysis is performed by preparing a working curve of intensity as a function of concentration of a series of standard solutions (i.e., solutions containing known concentrations of the component being analyzed). Working curves also are used with other analytical methods, including absorptiometry. The intensity of the scattered radiation in the analyte is measured and compared to the working curve. The concentration of the analyte corresponds to the concentration on the curve that has an intensity identical to that of the analyte.

For chemical analysis three forms of radiative scattering are important—namely, Tyndall, Raman, and Rayleigh scattering. Tyndall scattering occurs when the dimensions of the particles that are causing the scattering are larger than the wavelength of the scattered radiation. It is caused by reflection of the incident radiation from the surfaces of the particles, reflection from the interior walls of the particles, and refraction and diffraction of the radiation as it passes through the particles.

Raman and Rayleigh scattering occur when the dimensions of the scattering particles are less than 5 percent of the wavelength of the incident radiation. Both Rayleigh and Raman scattering are caused by the effect on the analyte of the fluctuating electromagnetic field that is associated with the passing incident radiation. The fluctuating field induces an electric dipole (separation of charges equal in size but opposite in sign) within the scattering particles that oscillates at the same frequency as the incident radiation. The oscillating dipole behaves as a point source of emitted radiation.

Turbidimetry and nephelometry

Scattered radiation can be used to perform quantitative analysis in either of two ways. If the apparatus is designed so that the detector is aligned with the cell and the radiative source, the detector responds to the decreased intensity of the incident radiation that is caused by scattering in the cell. Measurements of the decreased intensity are turbidimetric measurements; the technique is called turbidimetry. The measurements are completely analogous to absorption measurements. The only difference is in the phenomenon that causes the decreased radiative intensity. As with absorption measurements, the decreased intensity is related to the concentration of the scattering species in the cell at a constant wavelength. In both Tyndall scattering and Rayleigh scattering, the wavelength of the scattered radiation is identical to that of the incident radiation. Consequently, neither type provides information that is useful for qualitative analysis.

If the intensity of the scattered radiation is measured, rather than the decrease in intensity of the incident radiation, the method is known as nephelometry. The apparatus used for nephelometric measurements differs from that used for turbidimetric measurements in the placement of the detector. In nephelometry the detector is not aligned with the radiation source and the cell; normally it is placed perpendicular to the path of the incident radiation. Placing the detector out of the path of the incident radiation eliminates the possibility of measuring its intensity. Both nephelometry and turbidimetry are used with Tyndall scattering to quantitatively assay turbid solutions.

As mentioned above, Raman and Rayleigh scattering are caused by induced dipoles that are formed as the electromagnetic radiation passes the scattering particles. Raman scattering differs from Rayleigh scattering in that in the former the induced dipole relaxes to a different vibrational level than it originally had. Accordingly, the wavelength of the scattered radiation differs from the wavelength of the incident radiation by an amount corresponding to the difference between the particle's original and final vibrational levels. Shifts between the wavelengths of the incident radiation and the scattered radiation correspond to differences in vibrational levels within the scattering molecule and therefore can be used for qualitative analysis in much the same way that infrared spectrophotometry is used.

Refractometry

Another category of spectral analysis in which the incident radiation changes direction is refractometry. The refractive index of a substance is defined as the ratio of the velocity of electromagnetic radiation in a vacuum to its velocity in the medium of interest. Because it is difficult to accurately measure velocities as large as those of electromagnetic radiation, the refractive index is determined from the extent to which the radiation changes direction, owing to the decrease in velocity, as it passes from one medium into another. This phenomenon is refraction. Measurements of refractive index are used to qualitatively analyze pure substances because each substance has a constant and unique refractive index that can be determined with great accuracy. Quantitative analysis of simple mixtures containing known components is possible because the refractive index changes with the composition of the mixture.

Emitted Radiation

The spectroanalytical methods in the final major category utilize measurements of emitted radiation. Except for a few radionuclides that spontaneously emit radiation, emission occurs only after initial excitation of the analyte by an external source of energy.

Luminescence

In the most common case excitation occurs after the absorption of electromagnetic radiation. The absorption process is identical to that which occurs during absorptiometric measurements. After ultraviolet-visible absorption, an electron in the analyte molecule or atom resides in an upper electron orbital with one or more vacant orbitals nearer to the nucleus. Emission occurs when the excited electron returns to a lower electron orbital. The emitted radiation is termed luminescence. Luminescence is observed at energies that are equal to or less than the energy corresponding to the absorbed radiation.

After initial absorption, emission can occur by either of two mechanisms. In the most common form of luminescence, the excited electron returns to the lower electron orbital without inverting its spin—i.e., without changing the direction in which the electron rotates in the presence of a magnetic field. This phenomenon, known as fluorescence, occurs immediately after absorption. When absorption ceases, fluorescence also immediately ceases.

Although it occurs with low probability, the excited electron sometimes returns to a lower electron orbital by a path in which the electron first inverts its spin while moving to a slightly lower energy state and then inverts the spin again while returning to the original spin state in the unexcited electron orbital. Emission of ultraviolet-visible radiation occurs during the transition from the excited, inverted spin state to the unexcited electron orbital. Because inversion of the spinning electron during the last transition can require a relatively long time, the emission does not immediately cease when the absorption ceases. The resulting luminescence is called phosphorescence. Both fluorescence and phosphorescence can be used for analysis. Fluorescence can be distinguished from phosphorescence by the time delay in emission that occurs during the latter. If the luminescence immediately stops when the exciting radiation is cut off, it is fluorescence; if the luminescence continues, it is phosphorescence.

Owing to the arrangement of electron orbitals in molecules and atoms, phosphorescence is observed only in polyatomic species, whereas fluorescence can be observed in atoms as well as in polyatomic species. When fluorescence is observed in discrete, gaseous atoms, it is termed atomic fluorescence.

The apparatus used to make fluorescent and phosphorescent measurements is similar to that used to make measurements of scattered radiation. The detector is usually placed perpendicular to the path of the incident radiation in order to eliminate the possibility of monitoring the incident radiation. Devices that are used to measure fluorescence are fluorometers, and those that are employed to measure phosphorescence are phosphorimeters. Phosphorimeters differ from fluorometers in that they monitor luminescent intensity while the exciting radiation is not striking the cell.

At dilute concentrations, the intensity of the luminesced radiation is directly proportional to the concentration of the emitting species. As with other spectral methods, qualitative analysis is performed by comparing the spectrum of the analyte (a plot of the intensity of emitted radiation as a function of wavelength) with spectra of known substances.

Luminescence can be initiated by a process other than absorption of electromagnetic radiation. Some atoms can be sufficiently excited to emit radiation when exposed to the heat in a flame. The analytical technique that measures the wavelength and/or the intensity of emitted radiation from a flame is flame emission spectrometry. If electrical energy in the form of a spark or an arc is used to excite the analyte prior to measuring the intensity of emitted radiation, the method is atomic emission spectrometry. If a chemical reaction is used to initiate the luminescence, the technique is chemiluminescence; if an electrochemical reaction causes the luminescence, it is electrochemiluminescence.

X-ray Emission

X-ray emission spectrometry is the group of analytical methods in which emitted X-ray radiation is monitored. X rays are emitted when an electron in an outer orbital falls into a vacancy in an inner orbital. The vacancy is created by bombarding the atom with electrons, protons, alpha particles, or another type of particles. The vacancy also can be created by absorption of X-ray radiation or by nuclear capture of an inner-shell electron as it approaches the nucleus. Often the bombardment is sufficiently energetic to cause the inner orbital electron to be completely removed from the atom, thereby forming an ion with a vacant inner orbital.

Emitted X rays are used for qualitative and quantitative analysis in much the same way that emitted ultraviolet-visible radiation is employed in fluorometry. X-ray fluorescence is used more often for chemical analysis than the other X-ray methods. The diffraction pattern of X rays that are passed through solid crystalline materials is useful for determining the crystalline structure of solids. The analytical method that measures the diffraction patterns for the purpose of determining structure is termed X-ray diffraction analysis.

Several methods of surface analysis utilize X rays. Particle-induced X-ray emission (PIXE) is the method in which a small area on the surface of a sample is bombarded with accelerated particles and the resulting fluoresced X rays are monitored. If the bombarding particles are protons and the analytical technique is used to obtain an elemental map of a surface, the apparatus utilized is a proton microprobe. An electron microprobe functions in much the same manner. The scanning electron microscope utilizes electrons to bombard a surface, but the intensity of either backscattered (deflected through angles greater than 90°) or transmitted electrons is measured rather than the intensity of X rays. Electron microscopes are often used in conjunction with X-ray spectrometers to obtain information about surfaces.

Electron spectroscopy

Electron spectroscopy comprises a group of analytical methods that measure the kinetic energy of expelled electrons after initial bombardment of the analyte with X rays, ultraviolet radiation, ions, or electrons. When X rays are used for the bombardment, the analytical method is called either electron spectroscopy for chemical analysis (ESCA) or X-ray photoelectron spectroscopy (XPS). If the incident radiation is ultraviolet radiation, the method is termed ultraviolet photoelectron spectroscopy (UPS) or photoelectron spectroscopy (PES). When the bombarding particles are electrons and different emitted electrons are monitored, the method is Auger electron spectroscopy (AES). Other forms of less frequently used electron spectroscopy are available as well.

Radiochemical methods

During use of the radiochemical methods, spontaneous emissions of particles or electromagnetic radiation from unstable atomic nuclei are monitored. The intensity of the emitted particles or electromagnetic radiation is used for quantitative analysis, and the energy of the emissions is used for qualitative analysis. Emissions of alpha particles, electrons, positrons, neutrons, protons, and gamma rays can be useful. Gamma rays are energetically identical to X rays; however, they are emitted as a result of nuclear transformations rather than electron orbital transitions.

A radioisotope is an isotope of an element that spontaneously emits particles or radiation. Radioisotopes can be assayed using a radioanalytical method. In other cases, it is possible to bombard a nonradioactive sample with a particle or with radiation in order to transform temporarily all or part of the sample into a radioactive material that can be assayed. Sometimes it is possible to dilute a sample with a radioactive isotope of the assayed element. If the amount of the dilution can be deduced, the intensity of the emissions from the added radioisotope can be used to assay the nonradioactive analyte. This method is called isotopic dilution analysis.

Electroanalysis

The second major category of instrumental analysis is electroanalysis. The electroanalytical methods use electrically conductive probes, called electrodes, to make electrical contact with the analyte solution. The electrodes are used in conjunction with electric or electronic devices to which they are attached to measure an electrical parameter of the solution. The measured parameter is related to the identity of the analyte or to the quantity of the analyte in the solution.

The electroanalytical methods are divided into categories according to the electric parameters that are measured. The major electroanalytical methods include potentiometry, amperometry, conductometry, electrogravimetry, voltammetry (and polarography), and coulometry. The names of the methods reflect the measured electric property or its units. Potentiometry measures electric potential (or voltage) while maintaining a constant (normally nearly zero) electric current between the electrodes. Amperometry monitors electric current (amperes) while keeping the potential constant. Conductometry measures conductance (the ability of a solution to carry an electric current) while a constant alternating-current (AC) potential is maintained between the electrodes. Electrogravimetry is a gravimetric technique similar to the classical gravimetric methods that were described above, in which the solid that is weighed is deposited on one of the electrodes. Voltammetry is a technique in which the potential is varied in a regular manner while the current is monitored. Polarography is a subtype of voltammetry that utilizes a liquid metal electrode. Coulometry is a method that monitors the quantity of electricity (coulombs) that are consumed during an electrochemical reaction involving the analyte.

Most of the electroanalytical methods rely on the flow of electrons between one or more of the electrodes and the analyte. The analyte must be capable of either accepting one or more electrons (known as reduction) from the electrode or donating one or more electrons (oxidation) to the electrode. As an example, ferric iron (Fe3+) can be assayed because it can undergo a reduction to ferrous iron (Fe2+) by accepting an electron from the electrode as shown in the following reaction:

CONDUCTOMETRY

This is the method in which the capability of the analyte to conduct an electrical current is monitored. From Ohm's law (E = IR) it is apparent that the electric current (I) is inversely proportional to the resistance (R), where E represents potential difference. The inverse of the resistance is the conductance (G = 1/R). As the conductance of a solution increases, its ability to conduct an electric current increases.

In liquid solutions current is conducted between the electrodes by dissolved ions. The conductance of a solution depends on the number and types of ions in the solution. Generally small ions and highly charged ions conduct current better than large ions and ions with a small charge. The size of the ions is important because it determines the speed with which the ions can travel through the solution. Small ions can move more rapidly than larger ones. The charge is significant because it determines the amount of electrostatic attraction between the electrode and the ions.

Because conductometric measurements require the presence of ions, conductometry is not useful for the analysis of undissociated molecules. The measured conductance is the total conductance of all the ions in the solution. Since all ions contribute to the conductivity of a solution, the method is not particularly useful for qualitative analysis—i.e., the method is not selective. The two major uses of conductometry are to monitor the total conductance of a solution and to determine the end points of titrations that involve ions. Conductivity meters are used in conjunction with water purification systems, such as stills or deionizers, to indicate the presence or absence of ion-free water.

Conductometric titration curves are prepared by plotting the conductance as a function of the volume of added titrant. The curves consist of linear regions prior to and after the end point. The two linear portions are extrapolated to their point of intersection at the end point. As in other titrations, the end-point volume is used to calculate the amount or concentration of analyte that was originally present.

Voltammetry

Voltammetry can be used for both qualitative and quantitative analysis of a wide variety of molecular and ionic materials. In this method, a set of two or three electrodes is dipped into the analyte solution, and a regularly varying potential is applied to the indicator electrode relative to the reference electrode. The analyte electrochemically reacts at the indicator electrode. The reference electrode is constructed so that its potential is constant regardless of the solution into which it is dipped. Usually a third electrode (an auxiliary or counter electrode) is placed in the solution for the purpose of carrying most of the current. The potential is controlled between the indicator electrode and the reference electrode, but the current flows between the auxiliary electrode and the indicator electrode.

Classic polarography

The several forms of voltammetry differ in the type of varying potential that is applied to the indicator electrode. Polarography is voltammetry in which the indicator electrode is made of mercury or, rarely, another liquid metal. In classic polarography, mercury drops from a capillary tube. The surface of the mercury drop is the site of the electrochemical reaction with the analyte. The manner in which the direct-current (DC) potential of the indicator electrode varies with time is a potential (or voltage) ramp. In the most common case, the potential varies linearly with time, and the analytical method is known as linear sweep voltammetry (LSV).

Typically the potential is initially adjusted to a value at which no electrochemical reaction occurs at the indicator electrode. The potential is scanned in a direction that makes an electrochemical reaction more favourable. If reduction reactions are studied, the electrode is made more cathodic (negative); if oxidations are studied, the electrode is made more anodic (positive). Initially the current that is measured, before the electrochemical reaction begins, is small. As the electrode potential is changed, however, sufficient energy is applied to the indicator electrode to cause the reaction to take place. As the reaction occurs, electrons are withdrawn from the electrode (for electrochemical reductions) or donated to the electrode (for oxidations), and a current flows in the external electrical circuit. A voltammogram is a plot of the current as a function of the applied potential. The shape of a voltammogram depends on the type of indicator electrode and the potential ramp that are used.

This technique can be used for qualitative analysis because substances exhibit characteristic peaks or waves at different potentials. The height (current) of the wave or the peak, as measured by extrapolating the linear portion of the curve prior to the wave or peak and taking the difference between this extrapolated line and the current peak or plateau, is directly proportional to the concentration of the analyte and can be used for quantitative analysis. Normally the concentration corresponding to the peak or wave height of the analyte is determined from a working curve.

TRIANGULAR WAVE VOLTAMMETRY

Triangular wave voltammetry (TWV) is a method in which the potential is linearly scanned to a value past the potential at which an electrochemical reaction occurs and is then immediately scanned back to its original potential. A triangular wave voltammogram usually has a current peak on the forward scan and a second, inverted peak on the reverse scan representing the opposite reaction (oxidation or reduction) to that observed on the forward scan. Cyclic voltammetry is identical to TWV except in having more than one cycle of forward and reverse scans successively completed.

AC voltametry

During AC voltammetry an alternating potential is added to the DC potential ramp used for LSV. Only the AC portion of the total current is measured and plotted as a function of the DC potential portion of the potential ramp. Because flow of an alternating current requires the electrochemical reaction to occur in the forward and reverse directions, AC voltammetry is particularly useful for studying the extent to which electrochemical reactions are reversible.

Pulse and differential pulse voltammetry

Differential pulse voltammetry adds a periodically applied potential pulse (temporary increase in potential) to the voltage ramp used for LSV. The current is measured just prior to application of the pulse and at the end of the applied pulse. The difference between the two currents is plotted as a function of the LSV ramp potential. Pulse voltammetry utilizes a regularly increasing pulse height that is applied at periodic intervals. In pulse and differential pulse polarography the pulses are applied just before the mercury drop falls from the electrode. Typically the pulse is applied for about 50–60 milliseconds; and the current is measured during the last 17 milliseconds of each pulse. The voltammogram is a plot of the measured current as a function of the potential of the pulse. Many other variations of voltammetry also are available but are not as commonly used.

Electrogravimetry

Electrogravimetry was briefly described above as an interference removal technique. This method employs two or three electrodes, just as in voltammetry. Either a constant current or a constant potential is applied to the preweighed working electrode. The working electrode corresponds to the indicator electrode in voltammetry and most other electroanalytical methods. A solid product of the electrochemical reaction of the analyte coats the electrode during application of the electric current or potential. After the assayed substance has been completely removed from the solution by the electrochemical reaction, the working electrode is removed, rinsed, dried, and weighed. The increased mass of the electrode due to the presence of the reaction product is used to calculate the initial concentration of the analyte.

Assays done by using constant-current electrogravimetry can be completed more rapidly (typically 30 minutes per assay) than assays done by using constant-potential electrogravimetry (typically one hour per assay), but the constant-current assays are subject to more interferences. If only one component in the solution can react to form a deposit on the electrode, constant-current electrogravimetry is the preferred method. In constant-potential electrogravimetry the potential at the working electrode is controlled so that only a single electrochemical reaction can occur. The applied potential corresponds to the potential on the plateau of a voltammetric wave of the assayed material.

Coulometry

This technique is similar to electrogravimetry in that it can be used in the constant-current or in the constant-potential modes. It differs from electrogravimetry, however, in that the total quantity of electricity (coulombs) required to cause the analyte to completely react is measured rather than the mass of the electrochemical reaction product. It is not necessary for the reaction product to deposit on the electrode in order to perform a coulometric assay; however, it is necessary that the current that flows through the electrode be ultimately used for a single electrochemical reaction. This requirement can be met in constant-current coulometry by using the current to perform a coulometric titration. In a coulometric titration, the current generates a titrant that chemically reacts with the analyte. By keeping the precursor to the titrant in excess, it is possible to ensure that all of the current is used to form the chemical reactant. Because the electrochemically formed titrant reacts completely with the analyte, it is possible to perform a quantitative analysis. Constant-potential coulometry is not subject to the effects of interferences, because the potential of the working electrode is controlled at a value at which only a single electrochemical reaction can occur.

Amperometry

During amperometric assays the potential of the indicator electrode is adjusted to a value on the plateau of the voltammetric wave, as during controlled-potential electrogravimetry and coulometry (see above). The current that flows between the indicator electrode and a second electrode in the solution is measured and related to the concentration of the analyte. Amperometry is commonly employed in two ways, both of which take advantage of the linear variation in current at constant potential with the concentration of an electroactive species. A working curve of current as a function of concentration of a series of standard solutions is prepared, and the concentration of the analyte is determined from the curve, or amperometry is used to locate the end point in an amperometric titration. An amperometric titration curve is a plot of current as a function of titrant volume. The shape of the curve varies depending on which chemical species (the titrant, the analyte, or the product of the reaction) is electroactive. In each case the curve consists of linear regions before and after the end point that are extrapolated to intersection at the end point.

Potentiometry

This is the method in which the potential between two electrodes is measured while the electric current (usually nearly zero) between the electrodes is controlled. In the most common forms of potentiometry, two different types of electrodes are used. The potential of the indicator electrode varies, depending on the concentration of the analyte, while the potential of the reference electrode is constant. Potentiometry is probably the most frequently used electroanalytical method. It can be divided into two categories on the basis of the nature of the indicator electrode. If the electrode is a metal or other conductive material that is chemically and physically inert when placed in the analyte, it reflects the potential of the bulk solution into which it is dipped. Electrode materials that are commonly used for this type of potentiometry include platinum, gold, silver, graphite, and glassy carbon.

INERT-INDICATOR-ELECTRODE POTENTIOMETRY

Inert-indicator-electrode potentiometry utilizes oxidation-reduction reactions. The potential of a solution that contains an oxidation-reduction couple (e.g., Fe3+ and Fe2+) is dependent on the identity of the couple and on the activities of the oxidized and reduced chemical species in the couple. For a general reduction half reaction of the form Ox + ne- '! Red, where Ox is the oxidized form of the chemical species, Red is the reduced form, and n is the number of electrons (e") transferred during the reaction, the potential can be calculated by using the Nernst equation (equation 2). In the Nernst equation E is the potential at the indicator electrode, E° is the standard potential of the electrochemical reduction (a value that changes as the chemical identity of the couple changes), R is the gas law constant, T is the absolute temperature of the solution, n is the number of electrons transferred in the reduction (the value in the half reaction), F is the faraday constant, and the aOx and aRed terms are the activities of the oxidized and reduced chemical species, respectively, in the solution. The activities can be replaced by concentrations of the ionic species if the solution is sufficiently dilute.

The most common use for potentiometry with inert-indicator electrodes is determining the end points of oxidation-reduction titrations. A potentiometric titration curve is a plot of potential as a function of the volume of added titrant. The curves have an "S" or backward "S" shape, where the end point of the titration corresponds to the inflection point.

Crystallography

Crystallography is the science that examines the arrangement of atoms in solids. The word "crystallography" derives from the Greek words crystallon = cold drop / frozen drop, with its meaning extending to all solids with some degree of transparency, and grapho = write.

Before the development of X-ray diffraction crystallography (see below), the study of crystals was based on their geometry. This involves measuring the angles of crystal faces relative to theoretical reference axes (crystallographic axes), and establishing the symmetry of the crystal in question. The former is carried out using a goniometer. The position in 3D space of each crystal face is plotted on a stereographic net, e.g. Wulff net or Lambert net. In fact, the pole to each face is plotted on the net. Each point is labelled with its Miller index. The final plot allows the symmetry of the crystal to be established.

Crystallographic methods now depend on the analysis of the diffraction patterns of a sample targeted by a beam of some type. Although X-rays are most commonly used, the beam is not always electromagnetic radiation. For some purposes electrons or neutrons are used. This is facilitated by the wave properties of the particles. Crystallographers often explicitly state the type of illumination used when referring to a method, as with the terms X-ray diffraction, neutron diffraction and electron diffraction.

These three types of radiation interact with the specimen in different ways. X-rays interact with the spatial distribution of the valence electrons, while electrons are charged particles and therefore feel the total charge distribution of both the atomic nuclei and the surrounding electrons. Neutrons are scattered by the atomic nuclei through the strong nuclear forces, but in addition, the magnetic moment of neutrons is non-zero. They are therefore also scattered by magnetic fields. When neutrons are scattered from hydrogen-containing materials, they produce diffraction patterns with high noise levels.

However, the material can sometimes be treated to substitute deuterium for hydrogen. Because of these different forms of interaction, the three types of radiation are suitable for different crystallographic studies.

Technique

Some materials studied using crystallography, proteins for example, do not occur naturally as crystals. Typically, such molecules are placed in solution and allowed to crystallize over days, weeks, or months through vapor diffusion. A drop of solution containing the molecule, buffer, and precipitants is sealed in a container with a reservoir containing a hygroscopic solution. Water in the drop diffuses to the reservoir, slowly increasing the concentration and allowing a crystal to form. If the concentration were to rise more quickly, the molecule would simply precipitate out of solution, resulting in disorderly granules rather than an orderly and hence usable crystal.

Once a crystal is obtained, data can be collected using a beam of radiation. Although many universities that engage in crystallographic research have their own X-ray producing equipment, synchrotrons are often used as X-ray sources, because of the purer and more complete patterns such sources can generate. Synchrotron sources also have a much higher intensity of X-ray beams, so data collection takes a fraction of the time normally necessary at weaker sources. Complementary neutron crystallography techniques are used to enhance hydrogen atoms positions. Such techniques are available in Neutron facilities.

Producing an image from a diffraction pattern requires sophisticated mathematics and often an iterative process of modelling and refinement. In this process, the mathematically predicted diffraction patterns of an hypothesized or "model" structure are compared to the actual pattern generated by the crystalline sample. Ideally, researchers make several initial guesses, which through refinement all converge on the same answer. Models are refined until their predicted patterns match to as great a degree as can be achieved without radical revision of the model. This is a painstaking process, made much easier today by computers.

The mathematical methods for the analysis of diffraction data only apply to patterns, which in turn result only when waves diffract from orderly arrays. Hence crystallography applies for the most part only to crystals, or to molecules which can be coaxed to crystallize for the sake of measurement. In spite of this, a certain amount of molecular information can be deduced from the patterns that are generated by fibers and powders, which while not as perfect as a solid crystal, may exhibit a degree of order. This level of order can be sufficient to deduce the structure of simple molecules, or to determine the coarse features of more complicated molecules. For example, the double-helical structure of DNA was deduced from an X-ray diffraction pattern that had been generated by a fibrous sample.

Crystallography in Materials Engineering

Crystallography is a tool that is often employed by materials scientists. In single crystals, the effects of the crystalline arrangement of atoms is often easy to see macroscopically, because the natural shapes of crystals reflect the atomic structure. In addition, physical properties are often controlled by crystalline defects. The understanding of crystal structures is an important prerequisite for understanding crystallographic defects. Mostly, materials do not occur in a single crystalline, but poly-crystalline form, such that the powder diffraction method plays a most important role in structural determination.

A number of other physical properties are linked to crystallography. For example, the minerals in clay form small, flat, platelike structures. Clay can be easily deformed because the platelike particles can slip along each other in the plane of the plates, yet remain strongly connected in the direction perpendicular to the plates. Such mechanisms can be studied by crystallographic texture measurements.

In another example, iron transforms from a body-centered cubic (bcc) structure to a face-centered cubic (fcc) structure called austenite when it is heated. The fcc structure is a close-packed structure, and the bcc structure is not, which explains why the volume of the iron decreases when this transformation occurs.

Crystallography is useful in phase identification. When performing any process on a material, it may be desired to find out what compounds and what phases are present in the material. Each phase has a characteristic arrangement of atoms. Techniques like X-ray or neutron diffraction can be used to identify which patterns are present in the material, and thus which compounds are present.

Crystallography covers the enumeration of the symmetry patterns which can be formed by atoms in a crystal and for this reason has a relation to group theory and geometry.

7

GRAVIMETRIC ANALYSIS

All Gravimetric analyses rely on some final determination of weight as a means of quantifying an analyte. Since weight can be measured with greater accuracy than almost any other fundamental property, gravimetric analysis is potentially one of the most accurate classes of analytical methods available. These methods are among the oldest of analytical techniques, and they may be lengthy and tedious. Samples may have to be extensively treated to remove interfering substances. As a result, only a very few gravimetric methods are currently used in environmental analysis.

There are four fundamental types of gravimetric analysis: physical gravimetry, thermogravimetry, precipitative gravimetric analysis, and electrodeposition. These differ in the preparation of the sample before weighing of the analyte. Physical gravimetry is the most common type used in environmental engineering. It involves the physical separation and classification of matter in environmental samples based on volatility and particle size (e.g., total suspended solids). With thermogravimetry, samples are heated and changes in sample mass are recorded. Volatile solids analysis is an important example of this type of gravimetric analysis. As the name implies, precipitative gravimetry relies on the chemical precipitation of an analyte. Its most important application in the environmental field is with the analysis of sulfite. Electrodeposition involves the electrochemical reduction of metal ions at a cathode, and simultaneous deposition of the ions on the cathode.

Gravimetric analysis describes a set of methods in analytical chemistry for the quantitative determination of an analyte based on the mass of a solid. A simple example is the measurement of solids suspended in a water sample: A known volume of water is filtered, and the collected solids are weighed. Gravimetric analysis is a technique through which the amount of an analyte (the ion being analyzed) can be determined through the measurement of mass. Gravimetric analyses depend on comparing the masses of two compounds containing the analyte. The principle behind gravimetric analysis is that the mass of an ion in a pure compound can be determined and then used to find the mass percent of the same ion in a known quantity of an impure compound. In order for the analysis to be accurate, certain conditions must be met:

1. The ion being analyzed must be completely precipitated.
2. The precipitate must be a pure compound.
3. The precipitate must be easily filtered.

An example of a gravimetric analysis is the determination of chloride in a compound. In order to do a gravimetric analysis, a cation must be found that forms an insoluble compound with chloride. This compound must also be pure and easily filtered. The solubility rules indicate that Ag+, Pb2+, and Hg22+ form insoluble chlorides. Therefore silver chloride could be used to determine % Cl-, because it is insoluble (that is, about 99.9% of the silver is converted to AgCl) and it can be formed pure and is easily filtered.

In most cases, the analyte must first be converted to a solid by precipitation with an appropriate reagent. The precipitate can then be collected by filtration, washed, dried to remove traces of moisture from the solution, and weighed. The amount of analyte in the original sample can then be calculated from the mass of the precipitate and its chemical composition.

In other cases, it may be easier to remove the analyte by vaporization. The analyte might be collected—perhaps in a cryogenic trap or on some absorbent material such as activated carbon — and measured directly. Or, the sample can be weighed before and after it is dried; the difference between the two masses gives the mass of analyte lost. This is especially useful in determining the water content of complex materials such as foodstuffs.

Gravimetric analysis, which by definition is based upon the measurement of mass, can be generalized into two types; precipitation and volatilization. The quantitative determination of a substance by the precipitation method of gravimetric analysis involves isolation of an ion in solution by a precipitation reaction, filtering, washing the precipitate free of contaminants, conversion of the precipitate to a product of known composition, and finally weighing the precipitate and determining its mass by difference. From the mass and known composition of the precipitate, the amount of the original ion can be determined.

For successful determinations the following criteria must be met: The desired substance must be completely precipitated. In most determinations the precipitate is of such low solubility that losses from dissolution are negligible. An additional factor is the "common ion" effect, this further reduces the solubility of the precipitate. When Cl- is precipitated out by addition of Ag+

Ag+ + Cl- AgCl(s)

The (low) solubility of AgCl is reduced still further by the excess of Ag+ which is added, pushing the equilibrium to the right. We can further decrease the solubility by decreasing the temperature of the solution by using an ice bath. The weighed form of the product should be of known composition. The product should be "pure" and easily filtered. It is usually difficult to obtain a product which is "pure", i.e. one which is free from impurities but careful precipitation and sufficient washing helps reduce the level of impurity.

Procedure

1. The sample is dissolved, if it is not already insoluble.
2. The solution may be treated to adjust the pH (so that the proper precipitate is formed, or to suppress the formation of other precipitates). If it is known that species are present which interfere (by also forming precipitates under the same conditions as the analyte), the sample might require treatment with a different reagent to remove these interferents.
3. The precipitating reagent is added at a concentration that favors the formation of a "good" precipitate (see below). This may require low concentration, extensive heating (often described as "digestion"), or careful control of the pH. Digestion can help reduce the amount of coprecipitation.
4. After the precipitate has formed and been allowed to "digest", the solution is carefully filtered. The filter is chosen to trap the precipitate; smaller particles are more difficult to filter.

o Depending on the procedure followed, the filter might be a piece of ashless filter paper in a fluted funnel, or a filter crucible. Filter paper is convenient because it does not typically require cleaning before use; however, filter paper can be chemically attacked by some solutions (such as concentrated acid or base), and may tear during the filtration of large volumes of solution.

o The alternative is a crucible whose bottom is made of some porous material, such as sintered glass, porcelain or sometimes metal. These are chemically inert and mechanically stable, even at elevated temperatures. However, they must be carefully cleaned to minimize contamination or carryover(cross-contamination). Crucibles are often used with a mat of glass or asbestos fibers to trap small particles.

- After the solution has been filtered, it should be tested to make sure that the analyte has been completely precipitated. This is easily done by adding a few drops of the precipitating reagent; if a precipitate is observed, the precipitation is incomplete.

5. After filtration, the precipitate – including the filter paper or crucible – is heated. This achieves three purposes:

- The remaining moisture is removed (drying).
- Secondly, the precipitate is converted to a more chemically stable form. For instance, calcium ion might be precipitated using oxalate ion, to produce calcium oxalate (CaC2O4); it might then be heated to convert it into the oxide (CaO). It is vital that the empirical formula of the weighed precipitate be known, and that the precipitate be pure; if two forms are present, the results will be inaccurate.
- The precipitate cannot be weighed with the necessary accuracy in place on the filter paper; nor can the precipitate be completely removed from the filter paper in order to weigh it. The precipitate can be carefully heated in a crucible until the filter paper has burned away; this leaves only the precipitate. (As the name suggests, "ashless" paper is used so that the precipitate is not contaminated with ash.)

6. After the precipitate is allowed to cool (preferably in a desiccator to keep it from absorbing moisture), it is weighed (in the crucible). The mass of the crucible is subtracted from the combined mass, giving the mass of the precipitated analyte. Since the composition of the precipitate is known, it is simple to calculate the mass of analyte in the original sample.

Advantages

Gravimetric analysis, if methods are followed carefully, provides for exceedingly precise analysis. In fact, gravimetric analysis was used to determine the atomic masses of many elements to six figure accuracy. Gravimetry provides very little room for instrumental error and does not require a series of standards for calculation of an unknown. Also, methods often do not require expensive equipment. Gravimetric analysis, due to its high degree of accuracy, when performed correctly, can also be used to calibrate other instruments in lieu of reference standards.

Disadvantages

Gravimetric analysis usually only provides for the analysis of a single element, or a limited group of elements, at a time. Comparing modern dynamic flash combustion coupled with gas chromatography with traditional combustion analysis will show that the former is both faster and allows for simultaneous determination of multiple elements while traditional determination allowed only for the determination of carbon and hydrogen. Methods are often convoluted and a slight mis-step in a procedure can often mean disaster for the analysis (colloid formation in precipitation gravimetry, for example). Compare this with hardy methods such as spectrophotometry and one will find that analysis by these methods is much more efficient.

Mechanism of Precipitation

After the addition of the precipitating agent to the solution of the ion under analysis there is an initial induction period before nucleation occurs. This induction period may range from a very short time period to one which is relatively long, ranging from almost instantaneous to several minutes.

After induction, nucleation occurs, here small aggregates or nuclei of atoms form and it is from these "clumps" of atoms that the crystals which form the filtrate will grow. These nuclei may be composed of just a few atoms each so there may be up to 1010 of the nuclei per mole of precipitating product. As these nuclei form ions from the solution (which at this point are in excess) congregate around them. For example if hydrochloric acid were added very slowly to a solution of silver nitrate, silver chloride nuclei would form and silver ions (which would be in excess relative to Cl- ions) would congregate around them.

In addition to the primary adsorbed silver ion, there are some nitrate ions aggregating further from the AgCl nucleus. These are counter ions and tend to aggregate around the [AgCl:Ag]+ center because these centers have a net positive charge (excess Ag+) and additional negative charge is required to maintain electrical neutrality. The counter ions are less tightly held than the primary adsorbed ions and the counter ion layer is somewhat diffuse and contains ions other than those of the counter ions. These layers of charges are known as the electric double layer.

After nucleation growth occurs, large nuclei grow at the expense of smaller nuclei which dissolve. This process helps produce more easily filtered crystals (since it produces larger crystals). Growth of larger nuclei or crystallites can be encouraged by digestion, a process which involves heating the solid and mother liquor for a certain period of time. During digestion, small particles dissolve and larger ones grow. Digestion of the product is an important practical process and you will find that most if not all gravimetric analysis involve a digestion period.

Conditions for analytical precipitation

In an ideal world, an analytical precipitate for gravimetric analysis should consist of perfect crystals large enough to be easily washed and filtered. The perfect crystal would be free from impurities and be large enough so that it presented a minimum surface area onto which foreign ions could be adsorbed. The precipitate should also be "insoluble" (i.e. be of such slight solubility that loses from dissolution would be minimal).

Without going into detail, it has been shown (Von Weimarn) that the particle size of precipitates is inversely proportional to the relative supersaturation of the solution during precipitation; relative supersaturation = (Q-S)/S

For the best possible results, conditions need to be adjusted such that Q will be as low as possible and S will be relatively large.

The following methods are used to approach these criteria, (we will employ the first);

Precipitation from hot solution. The solubility S of precipitates increases with temperature and so an increase in S decreases the supersaturation.

Precipitation from dilute solution. (We will employ this method next semester for the crystallization of a metal oxalate) This keeps Q low. Slow addition of precipitating reagent with effective stirring. This also keeps Q low, stirring prevents local high concentrations of the precipitating agent.

Precipitation at a pH near the acidic end of the pH range in which the precipitate is quantitative. Many precipitates are more soluble at the lower (more acidic) pH values and so the rate of precipitation is slower.

Digestion of the precipitate. (Also the digestion period results in some improvement in the internal perfection of the crystal structure [sometimes called ripening], here some internal foreign atoms may be expelled).

Coprecipitation

This is anything unwanted which precipitates with the thing you do want. Coprecipitation occurs to some degree in every gravimetric analysis (especially barium sulfate and those involving hydrous oxides). You cannot avoid it - all you can do is minimize it by careful precipitation and thorough washing.

Surface adsorption

Here unwanted material is adsorbed onto the surface of the precipitate. Digestion of a precipitate reduces the amount of surface area and hence the area available for surface adsorption. Washing can also remove surface material.

Occlusion

This is a type of coprecipitation in which impurities are trapped within the growing crystal.

Postprecipitation

Sometimes a precipitate standing in contact with the mother liquor becomes contaminated by the precipitation of an impurity on top of the desired precipitate.

Washing and Filtering

Problems with coprecipitation and surface adsorption may be reduced by careful washing of the precipitate. With many precipitates, peptization occurs during washing. Here part of the precipitate reverts to the colloidal form e.g.

AgCl(colloidal) AgCl(s)

This results in the loss of part of the precipitate because the colloidal form may pass through on filtration. By washing with ice cold water this can be minimized

Drying the solid

Generally the solids are dried at about 120oC but conditions for drying can vary considerably. To determine the correct drying regime, a thermogravimetric balance may be used.

Gravimetry

Gravimetry is the measurement of the strength of a gravitational field. Gravimetry may be used when either the magnitude of gravitational field or the properties of matter responsible for its creation are of interest. The term gravimetry or gravimetric is also used in chemistry to define a class of analytical procedures, called gravimetric analysis relying upon weighing a sample of material.

Units of measurement

Gravity is usually measured in units of acceleration. In the SI system of units, the standard unit of acceleration is 1 metre per second squared (abbreviated as m/s2). Other units include the gal (sometimes known as a galileo, in either case with symbol Gal), which equals 1 centimetre per second squared, and the g (gn), equal to 9.80665 m/s2. The value of the gn approximately equals the acceleration due to gravity at the Earth's surface (although the actual acceleration g varies fractionally from place to place).

How gravity is measured

An instrument used to measure gravity is known as a gravimeter, or gravitometer. Since general relativity regards the effects of gravity as indistinguishable from the effects of acceleration, gravimeters may be regarded as special purpose accelerometers. Many weighing scales may be regarded as simple gravimeters. In one common form, a spring is used to counteract the force of gravity pulling on an object. The change in length of the spring may be calibrated to the force required to balance the gravitational pull. The resulting measurement may be made in units of force (such as the newton), but is more commonly made in units of gals.

More sophisticated gravimeters are used when precise measurements are needed. When measuring the Earth's gravitational field, measurements are made to the precision of microgals to find density variations in the rocks making up the Earth. Several types of gravimeters exist for making these measurements, including some that are essentially refined versions of the spring scale described above. These measurements are used to define gravity anomalies.

Besides precision, stability is also an important property of a gravimeter, as it allows the monitoring of gravity changes. These changes can be the result of mass displacements inside the Earth, or of vertical movements of the Earth's crust on which measurements are being made: remember that gravity decreases 0.3 mGal for every metre of height. The study of gravity changes belongs to geodynamics.

The majority of modern gravimeters use specially-designed metal or quartz zero-length springs to support the test mass. Zero length springs do not follow Hooke's Law, instead they have a force proportional to their length. The special property of these springs is that the natural resonant period of oscillation of the spring-mass system can be made very long - approaching a thousand seconds. This detunes the test mass from most local vibration and mechanical noise, increasing the sensitivity and utility of the gravimeter. Quartz and metal springs are chosen for different reasons; quartz springs are less affected by magnetic and electric fields while metal springs have a much lower drift (elongation) with time. The test mass is sealed in an air-tight container so that tiny changes of barometric pressure from blowing wind and other weather do not change the buoyancy of the test mass in air.

Spring gravimeters are, in practice, relative instruments which measure the difference in gravity between different locations. A relative instrument also requires calibration by comparing instrument readings taken at locations with known complete or absolute values of gravity. Absolute gravimeters provide such measurements by determining the gravitational acceleration of a test mass in vacuum. A test mass is allowed to fall freely inside a vacuum chamber and its position is measured with a laser interferometer and timed with an atomic clock. The laser wavelength is known to ±0.025 ppb and the clock is stable to ±0.03 ppb as well. Great care must be taken to minimize the effects of perturbing forces such as residual air resistance (even in vacuum), vibration, and magnetic forces. Such instruments are capable of an accuracy of about two parts per billion or 0.002 mGal and reference their measurement to atomic standards of length and time. Their primary use is for calibrating relative instruments, monitoring crustal deformation, and in geophysical studies requiring high accuracy and stability. However, absolute instruments are somewhat larger and significantly more expensive than relative spring gravimeters, and are thus relatively rare.

Gravimeters have been designed to mount in vehicles, including aircraft, ships and submarines. These special gravimeters isolate acceleration from the movement of the vehicle, and subtract it from measurements. The acceleration of the vehicles is often hundreds or thousands of times stronger than the changes being measured. A gravimeter (the Lunar Surface Gravimeter) was also deployed on the surface of the moon during the Apollo 17 mission, but did not work due to a design error. A second device (the Traverse Gravimeter Experiment) functioned as anticipated.

MICROGRAVIMETRY

Microgravimetry is a rising and important branch developed on the foundation of classical gravimetry. Microgravity investigations are carried out in order to solve various problems of engineering geology, mainly location of voids and their monitoring. Very detailed measurements of high accuracy can indicate voids of any origin, provided the size and depth are large enough to produce gravity effect stronger than is the level of confidence of relevant gravity signal.

Gravimeter

A gravimeter is an instrument used in gravimetry for measuring the local gravitational field of the Earth. A gravimeter is a type of accelerometer, specialized for measuring the constant downward acceleration of gravity, which varies by about 0.5% over the surface of the Earth. Though the essential principle of design is the same as in other accelerometers, gravimeters are typically designed to be much more sensitive in order to measure very tiny fractional changes within the Earth's gravity of 1 g, caused by nearby geologic structures or the shape of the Earth and by temporal tidal variations. This sensitivity means that gravimeters are susceptible to extraneous vibrations including noise that tend to cause oscillatory accelerations. In practice this is counteracted by integral vibration isolation and signal processing. The constraints on temporal resolution are usually less for gravimeters, so that resolution can be increased by processing the output with a longer "time constant". Gravimeters display their measurements in units of gals, instead of ordinary units of acceleration.

Gravimeters are used for petroleum and mineral prospecting, seismology, geodesy, geophysical surveys and other geophysical research, and for metrology.

There are two types of gravimeters: relative and absolute. Absolute gravimeters measure the local gravity in absolute units, gals. Relative gravimeters compare the value of gravity at one point with another. They must be calibrated at a location where the gravity is known accurately, and then transported to the location where the gravity is to be measured. They measure the ratio of the gravity at the two points.

Absolute gravimeters

Absolute gravimeters, which nowadays are made compact so they too can be used in the field, work by directly measuring the acceleration of a mass during free fall in a vacuum, when the accelerometer is rigidly attached to the ground.

The mass includes a retroreflector and terminates one arm of a Michelson interferometer. By counting and timing the interference fringes, the acceleration of the mass can be measured. A more recent development is a "rise and fall" version that tosses the mass upward and measures both upward and downward motion. This allows cancellation of some measurement errors, however "rise and fall" gravimeters are not in common use. Absolute gravimeters are used in the calibration of relative gravimeters, surveying for gravity anomalies (voids), and for establishing the vertical control network.

Relative gravimeters

Most common relative gravimeters are spring-based. They are used in gravity surveys over large areas for establishing the figure of the geoid over those areas. A spring-based relative gravimeter is basically a weight on a spring, and by measuring the amount by which the weight stretches the spring, local gravity can be measured. However, the strength of the spring must be calibrated by placing the instrument in a location with a known gravitational acceleration.

The most accurate relative gravimeters are superconducting gravimeters, which operate by suspending a liquid helium cooled diamagnetic superconducting niobium sphere in an extremely stable magnetic field; the current required to generate the magnetic field that suspends the niobium sphere is proportional to the strength of the Earth's gravitational field. The superconducting gravimeter achieves sensitivities of one nanogal, one thousandth of one billionth (10-12) of the Earth surface gravity. In a demonstration of the sensitivity of the superconducting gravimeter, Virtanen (2006), describes how an instrument at Metsähovi, Finland, detected the gradual increase in surface gravity as workmen cleared snow from its laboratory roof.

Transportable relative gravimeters also exist; they employ an extremely stable inertial platform to compensate for the masking effects of motion and vibration, a difficult engineering feat. The first transportable relative gravimeters were, reportedly, a secret military technology developed in the 1950-60s as a navigational aid for nuclear submarines. Subsequently in the 1980s, transportable relative gravimeters were reverse engineered by the civilian sector for use on ship, then in air and finally satellite borne gravity surveys.

Common Procedures in Gravimetric Analysis

a. Drying to a Constant Weight

All solids have a certain affinity for water, and may absorb moisture from the laboratory air. Reagents that readily pick up water are termed hygroscopic. Those that absorb so much water that they will dissolve in it and form a concentrated solution are called deliquescent (e.g., sodium hydroxide, trichloroacetic acid). These types of substances will continually increase in weight while exposed to the air. For this reason, many types of laboratory procedures require that a sample be dried to a constant, reproducible weight (i.e., absorbed moisture removed to some standard, low level). This is especially important for the gravimetric methods. Generally, the sample is dried in a 103° C to 110° C oven for about 1 hour and allowed to cool to room temperature in a desiccator. It is then weighed, and heated again for about 30 minutes. The sample is cooled and weighed a second time. The procedure is repeated until successive weighings agree to within 0.3 mg.

b. Description and Use of the Analytical Balance

The analytical balance is the most accurate and precise instrument in an environmental laboratory. Objects of up to 100 grams may be weighed to 6 significant figures. Volumetric glassware is accurate to no more than 4 significant figures, and the accuracy of complex analytical methods rarely justifies more than 2 significant figures. Analytical balances are generally used for gravimetric analyses, and for the preparation of standard solutions. Less demanding tasks, such as the weighing of sodium phosphate for the preparation of a pH buffer, should be done using the less accurate and more robust, toploading balance.

Early analytical balances were entirely mechanical with two weighing pans, one for the chemical, and one for the counterweights. Now, most analytical balances are hybrid mechanical and electronic with a single pan for the substance to be weighed. These balances use the substitution method of weighing. That is, the counterweight is fixed (hidden within the balance), and removable weights are mechanically added or subtracted from the sample side of the lever and fulcrum

PHYSICAL GRAVIMETRY

1. Total, Dissolved and Suspended Solids

a. Definitions

Total solids (TS) is generally defined as all matter in a water or wastewater sample that is not water. Because solids are not a specific chemical compound, but rather a diverse collection of dissolved and particulate matter, their concentration cannot be determined in an unambiguous way. Instead, they must be defined by the procedure used to estimated their concentration. This is referred to as an operational definition. Depending on a number of factors, the operationally defined solids concentration may be greater than, less than, or about the same as the true solids concentration. The operational definition that has been adopted through years of use is: all matter that remains as residue upon evaporation and drying at 180oC for one hour.

Total solids may be differentiated according to size into total dissolved solids (TDS) and total suspended solids. Once again, this is an operational distinction, whereby all solids passing through filter paper of a certain pore size (e.g., 1.5 microns, Whatman #934AH) are called dissolved, and those retained are termed suspended. Therefore, the overall definition of total dissolved solids is: all matter that is not retained by a 934AH filter, and is not lost by evaporation and drying at 180oC for one hour.

b. Significance to Environmental Engineering

Most of the impurities in potable waters are in the dissolved state, principally as inorganic salts. Thus, the parameters, "total solids" and especially "total dissolved solids" are of primary importance here. Waters containing high concentrations of inorganic salts are not suitable as sources of drinking water, because such materials are often difficult to remove during treatment. Finished drinking waters containing more than 1000 mg/L TDS are generally considered unacceptable. Waters of this type may also be unsuitable for agricultural purposes due to the harmful effects of high ionic concentrations on plants. In most natural waters, the TDS (total dissolved solids) concentration correlates well with total hardness (i.e., [Ca] + [Mg]). This is useful in assessing the corrosivity of a water and the need for softening.

One may also use conductivity measurements as a means of estimating the TDS (total dissolved solids) concentration. This measurement may also correlate well with the ionic strength, I. Ionic strength is numerically defined as half the sum of the product of the concentrations (in moles/liter) of each of the constituent ions times the square of their charge.

Ionic strength is particularly important for predicting the chemical behavior of pollutants in water treatment and in natural systems. Russell has developed the following relationship for predicting ionic strength from conductivity measurements:

I ~ 1.6 x 10-5 x Specific Conductance (in mmho/cm)

In a similar vein, the TDS has been used to estimate the ionic strength of natural waters via the Langlier approximation:

I ~ 2.5 x 10-5(TDS) ,where the TDS is expressed in mg/L.

The total suspended solids (TSS) content of natural waters is of interest for the purpose of assessing particle bed load and transport. High concentrations of suspended matter may be detrimental to aquatic life. In theory, TSS could be used for assessing particle removals during water treatment. However, nearly always the concentration of colloidal particles in water is measured as turbidity since this latter technique is faster and more precise. Volatile solids determinations are rarely made on potable water samples since far more accurate and precise information on organic content can be obtained by TOC (Total Organic Carbon) analysis.

In the realm of municipal wastewater, suspended solids analysis is by far the most important gravimetric method. It is used to evaluate the strength of the raw wastewater as well as the overall efficiency of treatment. Furthermore, most WWTP's have effluent standards of 10 to 30 mg/L suspended solids which may be legally enforceable. In particular, the volatile suspended solids is used in activated sludge to estimate the standing biomass in mixed liquor. In this type of application, the additional time and expense associated with more direct organic measurements such as TOC is not warranted. Total and volatile solids analyses are used in the design and operation of sludge digestors, vacuum-filters and incinerators.

As was the case with municipal wastewater, suspended solids analysis is useful as a means of assessing the strength of industrial wastewaters and the efficiency of industrial wastewater treatment. Since some industrial wastes contain unusually high concentrations of inorganic ions, the total solids and conductivity tests can be useful for determining the amenability of such waters to anaerobic treatment.

2. Sampling

Sampling for Solids Analysis. Suspended solids present a special sampling problem. In order to avoid unrepresentative sampling or particle break-up, it is important that: (1) the fluid stream to be sampled is completely mixed at the point of sampling; and (2) the disturbance of the flow pattern of this stream is minimized, especially at the mouth of the sampling device. This means that sample intake velocities should be similar to the flow velocities at the point of mixing. Note that intake velocity is less important for smaller particles. Finally, the sampler intake should face into the stream flow (within 20 degrees).

The possible presence of settleable solids may also require that some type of depth-integrated sample be collected. If it can be demonstrated that mixing is sufficient at the point of sampling to produce a homogeneous cross-section of settleable solids, depth-integrated sampling should not be necessary. Laboratory subsampling (removing a representative aliquot from a sample in the laboratory) is best accomplished by shaking, then rapidly pouring a portion of the sample into a receiving vessel. Samples should be collected in clean glass or high density polyethylene. Analysis should be conducted as soon after sampling as possible. Highly alkaline waters will leach metals from glass over time. Special care should be taken with waters high in manganese or iron. If these water are in contact with air or dissolved oxygen for extended periods of time the dissolved metals may become oxidized and precipitate.

3. Procedures

Total Solids (Total Residue). Total solids is determined by the final weight of a dried sample (minus tare) divided by the original sample volume. Evaporating dishes of platinum, vycor or porcelain may be used. Platinum is preferred, because it is more inert than the other two, and can be heated to a constant weight more easily. However, platinum is very expensive, so vycor is often used. Porcelain is difficult to bring to a constant weight, and its use should be avoided. Space permitting, evaporating dishes should be stored in a desiccator so as to avoid the collection of dust and absorption of moisture while not in use. The precision of this method has been estimated to be ± 4 mg or ± 5%. However, settled wastewater may give better precision, on the order of ± 1 mg.

1. Preheat a 100 mL evaporating dish at 550± 50° C for 1 hour, cool in a in a drying oven or in the open air (protected from dust) for 15-20 minutes, bring to room temperature in a desiccator, and weigh. Repeat until a constant weight is achieved.
2. Measure 75 mL of sample or a volume sufficient to yield 200 mg TS, whichever is less. Add this to the preweighed dish and evaporate to dryness in a drying oven set at 98° C. Alternatively, a steam bath may be used.
3. Dry for an additional hour at 103-105° C.
4. Cool in a desiccator and weigh. As always, the sample should be re-heated in accordance with #3 and re-weighed to achieve a constant weight.

Dissolved Solids (Filtrable Residue). Dissolved solids may be determined directly by analysis of the filtered sample for total solids, or indirectly by determining the suspended solids and subtracting this value from the total solids. When using the direct method, final drying may be conducted at one of two temperatures. Although solids measurements are operationally defined, and are therefore inherently without bias, one can postulate a method accuracy based on the theoretical solids definition (all that is not water). Using this as a basis of comparison, the choice of drying temperatures can have an important effect on accuracy. At 103-105° C positive bias may result from physically occluded water and the retention of waters of crystalization. This is especially true for waters high in calcium, magnesium, chloride or sulfate. The problem of physically occluded water may be exacerbated by the presence of large amounts of residue (i.e., more than 200 mg). On the other hand, some conversion of bicarbonate to carbon dioxide and subsequent volatilization may cause a small amount of negative bias. At 180± 2° C occluded water is removed, but some water of crystallation may still remain, especially in the presence of sulfates. Furthermore, at this higher temperature additional volatilization of certain inorganic salts (e.g., carbonate, some chlorides and nitrates) will occur. Overall, drying to 180± 2° C seems to give results for TDS that are closest to what one would calculate from a complete chemical analysis. For this reason, it is included as an option for the determination of TDS. In general, either temperature may be used for waters of low color and low alkalinity. Hard waters, alkaline waters, or highly-colored waters should probably be dried at the higher temperature.

1. Prewash a 934AH glass fiber filter by rinsing three times with 20 mL of distilled water. Maintain suction until the filter is dry.
2. Analyze the filtrate in accordance with the total solids procedure. Final drying (1 hour period) may be conducted at either 103-105° C or 180± 2° C.

Suspended Solids (Non-filtrable Residue). Suspended solids is measured directly by drying and weighing the solids retained during filtration. This approach is much more accurate for most waters than the indirect method of subtracting dissolved solids from total solids. Whatman 934AH glass fiber filters (nominal pore size 1.4 microns) are most commonly used. These may be placed in small aluminum foil weighing pans. The filters may be identified by etching a number into the small tab on the weighing pans. Filters should be weighed, heated and dried while remaining in the aluminum pans. They should only be removed for actual filtering operations. They should always be handled with tongs, and one must be careful not to tear them or leave fragments on the filtration apparatus. Precision has been found to be ± 5 mg/L at low concentrations, increasing up to ± 20 mg/L at values of 200 mg/L or more.

1. Prewash a 934AH glass fiber filter by rinsing three times with 20 mL of distilled water. Maintain suction until the filter is dry.
2. Dry this filter at 103-105° C for 1 hour, and cool in a desiccator.
3. Weigh the filter, then pass a water sample of sufficient volume to yield 50-200 mg suspended solids through it. Smaller volumes will result in reduced accuracy.
4. Dry for at least one hour at 103-105° C.
5. Cool in a desiccator and weigh. As always, the sample should be re-heated in accordance with #4 and re-weighed to achieve a constant weight (a drop of no more than 0.5 mg).

THERMOGRAVIMETRY AND COMBUSTION ANALYSIS

Thermogravimetry and combustion analysis involve the heating of a sample to 500° C or more with the oxidation and/or volatilization of some of the sample constituents. Either the change of sample weight is determined (thermogravimetry), or the combustion gases are trapped and weighed (combustion analysis). The former method is used for volatile solids analysis in environmental engineering. The latter is used in many fields of science for the determination of total carbon and hydrogen in solids.

With thermogravimetric methods, it is especially important to return the sample to room temperature before weighing. Otherwise the differences in temperature will create convection currents around the balance pan, which will severely disrupt method accuracy. A steady increase in apparent weight while the sample is on the pan indicates a problem of this type. Large vessels and samples will require longer cooling times to dissipate their excess heat.

1. Volatile Solids and Fixed Solids

Fixed solids are those that remain as residue after ignition at 550° C for 15 minutes. The weight of material lost is called the volatile solids. Thus the total operational definition for volatile solids would be: all matter lost upon ignition at 550° C for 15 minutes, but not lost upon drying at 103-105° C for 1 hour. The portion lost upon ignition is generally assumed to be equivalent to the organic fraction. The portion remaining is considered the inorganic fraction. For waters of moderate to high hardness, most of this is calcium carbonate which decomposes only at temperatures exceeding 800° C. When igniting a filter with suspended matter, one must be especially careful of the temperature; above 600° C glass fiber filters begin to melt and can loose a significant amount of weight in 15 minutes.

Combining the fractionations resulting from ignition and filtration, one arrives at a total of 9 separate categories: total solids (TS), fixed solids, volatile solids, total dissolved solids (TDS), fixed dissolved solids, volatile dissolved solids, total suspended solids (TSS), fixed suspended solids, and volatile suspended solids (VSS). In practice, only four of these (TS, TDS, TSS, and VSS) are commonly used. When comparing fixed solids with inorganic content, one would expect positive bias from incomplete oxidation of organic matter, and negative bias from decomposition of certain inorganics. Ammonium salts may be lost during low temperature drying or upon ignition. Most others are stable under the conditions used for volatile solids determination with the exception of magnesium carbonate (equation 15.4). Volatile solids may be effected by these as well as loss of recalcitrant waters of crystallation (positive bias), and previous losses of organic matter to volatilization during low-temperature drying (negative bias). A modest interlaboratory study found an average standard deviation of 11 mg/L on a sample of 170 mg/L volatile solids.

$MgCO_3$ ————————> $MgO + CO_2$ (15.4)

Analysis of sludge, sediment, and soil for fixed and volatile solids presents additional problems. Both negative and positive biases can be acute, so greater care must be taken in ignition temperature and time. Again, ammonium compounds (often present in sludge in the form of ammonium bicarbonate) may be lost during low temperature drying and therefore should not introduce a bias in volatile solids analysis (equation 15.5). However, occluded water can be especially problematic. Standard practice is to use a 1-hour ignition time. With sludge, sediment and soil, nearly all of the solids are in the undissolved state. Therefore, only total, total volatile, and total fixed fractions are normally determined.

NH_4HCO_3 —————> $NH_3 + H_2O + CO_2$ (15.5)

1. Dry and weigh a vessel containing the solids to be analyzed. For volatile and fixed suspended solids analysis, the filter (with residue) prepared for suspended solids analysis and dried to a constant weight may be used. For volatile and fixed total (or dissolved) solids, the evaporating dish (with residue) prepared for total (or dissolved) solids analysis dried to a constant weight should be used.
2. Ignite the sample and vessel in a preheated muffle furnace set at 550± 50° C for 15-20 min (water & wastewater) or 1 hour (sludge, sediment & soil).
3. Cool for 15 minutes in the open air in an area protected from dust.
4. Place vessel in a desiccator for final cooling to room temperature and weigh. Due to the approximate nature of this test samples are not generally re-heated and dried to a constant weight.

2. Carbon & Hydrogen Analysis

Total carbon and hydrogen can be determined in a solid by combustion in the presence of pure, dry oxygen. The combustion gases are oxidized completely to CO2 and H2O with the aid of a transition metal catalyst. The first trap, phosphorus pentoxide (P2O5), efficiently removes all water formed. Recall that P2O5 is deliquescent, and an efficient desiccant. The second contains ascarite, which is sodium hydroxide-impregnated asbestos. This will trap all of the carbon dioxide. The third tube protects the other two from backflow of atmospheric water and carbon dioxide. After the sample is completely oxidized, the tubes are weighed to determine the increase. This type of analysis is sometimes conducted on dried sludges, sediments, soils, and extracted aquatic organic matter.

PRECIPITATIVE GRAVIMETRIC ANALYSIS

Precipitative gravimetric analysis requires that the substance to be weighed be readily removed by filtration. In order for a non-filtrable precipitate to form, it must be supersaturated with respect to its solubility product constant. However, if it is too far above the saturation limit, crystal nucleation may occur at a rate faster than crystal growth (the addition of molecules to a crystal nucleus, eventually forming a non-filtrable crystal). When this occurs, numerous tiny micro-crystals are formed rather than a few large ones. In the extreme case, micro-crystals may behave as colloids and pass through a fibrous filter. To avoid this, precipitating solutions may be heated. Because the solubility of most salts increases with increasing temperature, this treatment will lower the relative degree of supersaturation and slow the rate of nucleation. Also, one might add the precipitant slowly with rapid mixing to avoid the occurrence of locally high concentrations.

Precipitative gravimetry is often practiced at high ionic strengths. This is to reduce the electric double layer thickness (salting-out effect) of the slowly forming crystals. When this occurs, electrostatic repulsion between the crystal and its precipitating molecules is reduced. Crystal growth can then occur more rapidly.

It is very important that the precipitate be pure and have the correct stoichiometry. Coprecipitation occurs when an unwanted ion or molecule becomes trapped in the precipitate. This may be due to inclusion or occlusion. Inclusion is the term used of a single subsitution in the crystal lattice by an ion of similar size. Occlusion refers to the physical trapping of a large pocket of impurities within the crystal. One technique for minimizing these problems is to remove the mother liquor, re-dissolve the precipitate and then re-precipitate. The second time the mother liquor will contain fewer unwanted ions capable of coprecipitation.

1. Sulfate

The method of choice for sulfate in waters and wastewaters is the precipitative gravimetric procedure using barium. If Ba(+II) is added in excess under acidic conditions, $BaSO_4$ is precipitated quantitatively. The reaction is allowed to continue for 2 hours or more at 80-90oC. This is to encourage the formation of $BaSO_4$ crystals (non-filtrable) from the initially formed colloidal precipitate (partially filtrable). The precipitate is washed, and then dried at 800‘C for 1 hour. Low pH is needed to avoid the precipitation of $BaCO_3$ and $Ba_3(PO_4)_2$. Positive bias may result from acid-resistant insoluble matter such as silica, sulfites which may oxidize to sulfate, and nitrate and chloride which will associate with barium and co-precipitate to a small extent with the barium sulfate. Negative bias may result from the presence of certain heavy metals (e.g., Cr, Fe) which can form soluble complexes with sulfate.

$$Ba^{+2} + SO_4^{-2} = BaSO_4(ppt.)$$

Chloride may be determined by precipitation with silver. Interfering ions likely to form insoluble silver salts are the other halogens (bromide, iodide), cyanide, and reduced sulfur species (sulfite, sulfide, and thiosulfate). Fortunately, the reduced sulfur compounds can be pre-oxidized with hydrogen peroxide, and the others are rarely present at high concentrations. Although AgCl can be determined gravimetrically, the recommended procedure for water and wastewater is to use a volumetric procedure with chromate as an indicator.

Thermal analysis

Thermal analysis is a branch of materials science where the properties of materials are studied as they change with temperature. Several methods are commonly used – these are distinguished from one another by the property which is measured:

- Dielectric thermal analysis (DEA): dielectric permittivity and loss factor
- Differential thermal analysis (DTA): temperature difference
- Differential scanning calorimetry (DSC): heat difference
- Dilatometry (DIL): volume
- Dynamic mechanical analysis (DMA) : mechanical stiffness and damping
- Evolved gas analysis (EGA) : gaseous decomposition products
- Laser flash analysis (LFA): thermal diffusivity and thermal conductivity
- Thermogravimetric analysis (TGA): mass
- Thermomechanical analysis (TMA): dimension
- Thermo-optical analysis(TOA): optical properties

Simultaneous Thermal Analysis (STA) generally refers to the simultaneous application of Thermogravimetry (TGA) and differential scanning calorimetry (DSC) to one and the same sample in a single instrument. The test conditions are perfectly identical for the TGA and DSC signals (same atmosphere, gas flow rate, vapor pressure of the sample, heating rate, thermal contact to the sample crucible and sensor, radiation effect, etc.). The information gathered can even be enhanced by coupling the STA instrument to an Evolved Gas Analyzer (EGA) like Fourier transform infrared spectroscopy (FTIR) or mass spectrometry (MS).

Other, less-common, methods measure the sound or light emission from a sample, or the electrical discharge from a dielectric material, or the mechanical relaxation in a stressed specimen. The essence of all these techniques is that the sample's response is recorded as a function of temperature (and time).

It is usual to control the temperature in a predetermined way - either by a continuous increase or decrease in temperature at a constant rate (linear heating/cooling) or by carrying out a series of determinations at different temperatures (stepwise isothermal measurements). More advanced temperature profiles have been developed which use an oscillating (usually sine or square wave) heating rate (Modulated Temperature Thermal Analysis) or modify the heating rate in response to changes in the system's properties (Sample Controlled Thermal Analysis).

In addition to controlling the temperature of the sample, it is also important to control its environment (e.g. atmosphere). Measurements may be carried out in air or under an inert gas (e.g. nitrogen or helium). Reducing or reactive atmospheres have also been used and measurements are even carried out with the sample surrounded by water or other liquids. Inverse gas chromatography is a technique which studies the interaction of gases and vapours with a surface - measurements are often made at different temperatures so that these experiments can be considered to come under the auspices of Thermal Analysis.

Atomic force microscopy uses a fine stylus to map the topography and mechanical properties of surfaces to high spatial resolution. By controlling the temperature of the heated tip and/or the sample a form of spatially resolved thermal analysis can be carried out.

Thermal analysis is also often used as a term for the study of heat transfer through structures. Many of the basic engineering data for modelling such systems comes from measurements of heat capacity and thermal conductivity.

Thermal Analysis of Pharma Materials

DSC,TG/DTA and TG/DTA-IR are often used for characterisation of pharma materials. DSC, alone or in combination with hot-stage microscopy, is able to differentiate between different polymorphic structures and, by using different heating rates, can investigate the transformations which occur during the polymorphic transformation. By using appropriate heating rates, polymorphic purity can be determined, and can involve heating rates up to 750°C/min. TGA is often used to measure residual solvents and moisture, but can also be used to determine solubility of pharma materials in solvents. Analysis of pharma materials is probably the largest area of application for thermal analysis.

Thermal Analysis of Polymers

Polymers represent another large area in which thermal analysis finds strong applications. Thermoplastic polymers are commonly found in everyday packaging and household items, but for the analysis of the raw materials, effects of the many additive used (including stabilisers and colours) and fine-tuning of the moulding or extrusion processing used can be achieved by using DSC. An example is oxidation induction time (OIT) by DSC which can determine the amount of oxidation stabiliser present in a thermoplastic (usually a polyolefin) polymer material. Compositional analysis is often made using TGA, which can separate fillers, polymer resin and other additives. TGA can also give an indication of thermal stability and the effects of additives such as flame retardants

Thermal analysis of composite materials, such as carbon fibre composites or glass epoxy composites are often carried out using DMA or DMTA, which can measure the stiffness of materials by determining the modulus and damping (energy absorbing) properties of the material. Aerospace companies often employ these analysers in routine quality control to ensure that products being manufactured meet the required strength specifications. Formula 1 racing car manufacturers also have similar requirements! DSC is used to determine the curing properties of the resins used in composite materials, and can also confirm whether a resin can be cured and how much heat is evolved during that process. Application of predictive kinetics analysis can help to fine-tune manufacturing processes. Another example is that TGA can be used to measure the fibre content of composites by heating a sample to remove the resin by application of heat and then determining the mass remaining.

Thermal Analysis of Metals

Production of many metals (cast iron, grey iron, ductile iron, compacted graphite iron, 3000 series aluminium alloys, copper alloys, silver, and complex steels) are aided by a production technique also referred to as thermal analysis. A sample of liquid metal is removed from the furnace or ladle and poured into a sample cup with a thermocouple embedded in it. The temperature is then monitored, and the phase diagram arrests (liquidus, eutectic, and solidus) are noted. From this information chemical composition based on the phase diagram can be calculated, or the crystalline structure of the cast sample can be estimated. Strictly speaking these measurements are cooling curves and a form of sample controlled thermal analysis whereby the cooling rate of the sample is dependent on the cup material (usually bonded sand) and sample volume which is normally a constant due to the use of standard sized sample cups.

Advanced techniques use differential curves to locate endothermic inflection points such as gas holes, and shrinkage, or exothermic phases such as carbides, beta crystals, inter crystalline copper, magnesium silicide, iron phosphide's and other phases as they solidify. Detection limits seem to be around 0.01% to 0.03% of volume.

In addition, integration of the area between the zero curve and the first derivative is a measure of the specific heat of that part of the solidification which can lead to rough estimates of the percent volume of a phase. (Something has to be either known or assumed about the specific heat of the phase versus the overall specific heat.) In spite of this limitation, this method is better than estimates from two dimensional micro analysis, and a lot faster than chemical dissolution.

Thermal Analysis of Foods

Most foods are subjected to variations in their temperature during production, transport, storage, preparation and consumption, e.g., pasteurization, sterilization, evaporation, cooking, freezing, chilling, etc. Temperature changes cause alterations in the physical and chemical properties of food components which influence the overall properties of the final product, e.g., taste, appearance, texture and stability. Chemical reactions such as hydrolysis, oxidation or reduction may be promoted, or physical changes, such as evaporation, melting, crystallization, aggregation or gelation may occur. A better understanding of the influence of temperature on the properties of foods enables food manufacturers to optimize processing conditions and improve product quality. It is therefore important for food scientists to have analytical techniques to monitor the changes that occur in foods when their temperature varies. These techniques are often grouped under the general heading of thermal analysis. In principle, most analytical techniques can be used, or easily adapted, to monitor the temperature-dependent properties of foods, e.g., spectroscopic (NMR, UV-visible, IR spectroscopy, fluorescence), scattering (light, X-rays, neutrons), physical (mass, density, rheology, heat capacity) etc. Nevertheless, at present the term thermal analysis is usually reserved for a narrow range of techniques that measure changes in the physical properties of foods with temperature (TG/DTG, DTA,DSC and Transition temperature).

Thermal Analysis of Printed Circuit Boards (PCB)

Power dissipation is an important issue in present-day PCB design. Power dissipation will result in temperature difference and pose a thermal problem to a chip. In addition to the issue of reliability, excess heat will also negatively affect electrical performance and safety. The working temperature of an IC should therefore be kept below the maximum allowable limit of the worst case. In general, the temperatures of junction and ambient are 125 °C and 55 °C, respectively. The ever-shrinking chip size causes the heat to concentrate within a small area and leads to high power density. Furthermore, denser transistors gathering in a monolithic chip and higher operating frequency cause a worsening of the power dissipation. Removing the heat effectively becomes the critical issue to be resolved.

Dielectric thermal analysis

Dielectric Thermal Analysis (DETA), or Dielectric Analysis (DEA), is a materials science technique similar to dynamic mechanical analysis except that an oscillating electrical field is used instead of a mechanical force. For investigation of the curing behavior of thermosetting resin systems, composite materials, adhesives and paints, Dielectric Analysis (DEA) can be used in accordance with ASTM E 2038 or E 2039. The great advantage of DEA is that it can be employed not only on a laboratory scale, but also in process.

Measuring Principle

In a typical test, the sample is placed in contact with two electrodes (the dielectric sensor) and a sinusoidal voltage (the excitation) is applied to one electrode. The resulting sinusoidal current (the response) is measured at the second electrode. The response signal is attenuated in amplitude and shifted in phase in relation to the mobility of the ions and alignment of the dipoles. Dipoles in the material will attempt to align with the electric field and ions (present as impurities) will move toward the electrode of opposite polarity. The dielectric properties of permittivity å' and loss factor å" are then calculated from this measured amplitude and phase change.

Differential thermal analysis

Differential thermal analysis (or DTA) is a thermoanalytic technique, similar to differential scanning calorimetry. In DTA, the material under study and an inert reference are made to undergo identical thermal cycles, while recording any temperature difference between sample and reference. This differential temperature is then plotted against time, or against temperature (DTA curve or thermogram). Changes in the sample, either exothermic or endothermic, can be detected relative to the inert reference. Thus, a DTA curve provides data on the transformations that have occurred, such as glass transitions, crystallization, melting and sublimation. The area under a DTA peak is the enthalpy change and is not affected by the heat capacity of the sample.

A DTA consists of a sample holder comprising thermocouples, sample containers and a ceramic or metallic block; a furnace; a temperature programmer; and a recording system. The key feature is the existence of two thermocouples connected to a voltmeter. One thermocouple is placed in an inert material such as Al2O3, while the other is placed in a sample of the material under study. As the temperature is increased, there will be a brief deflection of the voltmeter if the sample is undergoing a phase transition. This occurs because the input of heat will raise the temperature of the inert substance, but be incorporated as latent heat in the material changing phase.

In today's market most manufactures no longer make a true DTA but rather have incorporated this technology into a Thermogravimetric analysis (TGA), which provides both mass loss and thermal information. With today's advancements in software, even these instruments are being replaced by true TGA-DSC instruments that can provide the temperature and heat flow of the sample, simultaneously with mass loss.

A DTA curve can be used only as a finger print for identification purposes but usually the applications of this method are the determination of phase diagrams, heat change measurements and decomposition in various atmospheres.

DTA is widely used in the pharmaceutical and food industries. DTA may be used in cement chemistry, mineralogical research and in environmental studies. DTA curves may also be used to date bone remains or to study archaeological materials.

DIFFERENTIAL SCANNING CALORIMETRY

Differential scanning calorimetry or DSC is a thermoanalytical technique in which the difference in the amount of heat required to increase the temperature of a sample and reference is measured as a function of temperature. Both the sample and reference are maintained at nearly the same temperature throughout the experiment. Generally, the temperature program for a DSC analysis is designed such that the sample holder temperature increases linearly as a function of time. The reference sample should have a well-defined heat capacity over the range of temperatures to be scanned.

The technique was developed by E.S. Watson and M.J. O'Neill in 1962, and introduced commercially at the 1963 Pittsburgh Conference on Analytical Chemistry and Applied Spectroscopy. The first adiabatic differential scanning calorimeter that could be used in biochemistry was developed by P.L. Privalov and D.R. Monaselidze in 1964. The term DSC was coined to describe this instrument which measures energy directly and allows precise measurements of heat capacity.

The basic principle underlying this technique is that when the sample undergoes a physical transformation such as phase transitions, more or less heat will need to flow to it than the reference to maintain both at the same temperature. Whether less or more heat must flow to the sample depends on whether the process is exothermic or endothermic. For example, as a solid sample melts to a liquid it will require more heat flowing to the sample to increase its temperature at the same rate as the reference. This is due to the absorption of heat by the sample as it undergoes the endothermic phase transition from solid to liquid. Likewise, as the sample undergoes exothermic processes (such as crystallization) less heat is required to raise the sample temperature. By observing the difference in heat flow between the sample and reference, differential scanning calorimeters are able to measure the amount of heat absorbed or released during such transitions. DSC may also be used to observe more subtle physical changes, such as glass transitions. It is widely used in industrial settings as a quality control instrument due to its applicability in evaluating sample purity and for studying polymer curing.

DTA

An alternative technique, which shares much in common with DSC, is differential thermal analysis (DTA). In this technique it is the heat flow to the sample and reference that remains the same rather than the temperature. When the sample and reference are heated identically phase changes and other thermal processes cause a difference in temperature between the sample and reference. Both DSC and DTA provide similar information. DSC measures the energy required to keep both the reference and the sample at the same temperature whereas DTA measures the difference in temperature between the sample and the reference when they are both put under the same heat.

Dilatometer

A dilatometer is a scientific instrument that measures volume changes caused by a physical or chemical process. A familiar application of a dilatometer is the mercury-in-glass thermometer, in which the change in volume of the liquid column is read from a graduated scale. Because mercury has a fairly constant rate of expansion over normal temperature ranges, the volume changes are directly related to temperature.

Dilatometers have been used in the fabrication of metallic alloys, compressed and sintered refractory compounds, glasses, ceramic products, composite materials, and plastics.

Dilatometry is also used to monitor the progress of chemical reactions, particularly those displaying a substantial molar volume change (e.g., polymerisation). A specific example is the rate of phase changes.

There are a number of dilatometer types:

- Capacity dilatometers possess a parallel plate capacitor with a one stationary plate, and one moveable plate. When the sample length changes, it moves the moveable plate, which changes the gap between the plates. The capacitance is inversely proportional to the gap. Changes in length of 10 picometres can be detected.

- Connecting rod (push rod) dilatometer, the sample which can be examined is in the furnace. A connecting rod transfers the thermal expansion to a strain gauge, which measures the shift. Since the measuring system (connecting rod) is exposed to the same temperature as the sample and thereby likewise expands, one obtains a relative value, which must be converted afterwards. Matched low-expansion materials and differential constructions can be used to minimize the influence of connecting rod expansion

- High Resolution - Laser Dilatometer Highest resolution and absolute accuracy is possible with a Michelson Interferometer type Laser Dilatometer. Resolution goes up to picometres. On top the principle of interference measurement give the possibility for much higher accuracy's and it is an absolute measurement technique with no need of calibration.
- Optical dilatometer is an instrument that measures dimension variations of a specimen heated at temperatures that generally range from 25 to 1400 °C. The optical dilatometer allows the monitoring of materials' expansions and contractions by using a non-contact method: optical group connected to a digital camera captures the images of the expanding/contracting specimen as function of the temperature with a resolution of about ±70 micrometre per pixel. As the system allows to heat up the material and measures its longitudinal/vertical movements without any contact between instrument and specimen, it is possible to analyse the most ductile materials, such as the polymers, as well as the most fragile, such as the incoherent ceramic powders for sintering process.

For simpler measurements in a temperature range from 0 to 100 °C, where water is heated up and flow or over the sample. If linear coefficients of expansion of a metal is to be measured, hot water will running through a pipe made from the metal. The pipe warms up to the temperature of the water and the relative expansion can be determined as a function of the water temperature.

For the measurement of the volumetric expansion of liquids one takes a large glass container filled with water. In an expansion tank (glass container with an accurate volume scale) with the sample liquid. If one heats the water up, the sample liquid expands and the volume changes is read. However the expansion of the sample container must also be taken into consideration.

The expansion and retraction coefficient of gases cannot be measured using dilatometer, since the pressure plays a role here. For such measurements a gas thermometer is more suitable.

Dilatometers often include a mechanism for controlling temperature. This may be a furnace for measurements at elevated temperatures (temperatures to 2000 °C), or a cryostat for measurements at temperatures below room temperature. Metallurgical applications often involve sophisticated temperature controls capable of applying precise temperature-time profiles for heating and quenching the sample.

INSTRUMENTATION

The instrumentation of a DMA consists of a displacement sensor such as a linear variable differential transformer, which measures a change in voltage as a result of the instrument probe moving through a magnetic core, a temperature control system or furnace, a drive motor (a linear motor for probe loading which provides load for the applied force), a drive shaft support and guidance system to act as a guide for the force from the motor to the sample, and sample clamps in order to hold the sample being tested. Depending on what is being measured, samples will be prepared and handled differently.

Types of analyzers

There are two main types of DMA analyzers used currently: forced resonance analyzers and free resonance analyzers. Free resonance analyzers measure the free oscillations of damping of the sample being tested by suspending and swinging the sample. A restriction to free resonance analyzers is that it is limited to rod or rectangular shaped samples, but samples that can be woven/braided are also applicable. Forced resonance analyzers are the more common type of analyzers available in instrumentation today. These types of analyzers force the sample to oscillate at a certain frequency and are reliable for performing a temperature sweep.

Analyzers are made for both stress (force) and strain (displacement) control. In strain control, the probe is displaced and the resulting stress of the sample is measured by implementing a force balance transducer, which utilizes different shafts. The advantages of strain control include a better short time response for materials of low viscosity and experiments of stress relaxation are done with relative ease. In stress control, a set force is applied to the same and several other experimental conditions (temperature, frequency, or time) can be varied. Stress control is typically less expensive than strain control because only one shaft is needed, but this also makes it harder to use. Some advantages of stress control include the fact that the structure of the sample is less likely to be destroyed and longer relaxation times/ longer creep studies can be done with much more ease. Characterizing low viscous materials come at a disadvantage of short time responses that are limited by inertia. Stress and strain control analyzers give about the same results as long as characterization is within the linear region of the polymer in question. However, stress control lends a more realistic response because polymers have a tendency to resist a load.

Stress and strain can be applied via torsional or axial analyzers. Torsional analyzers are mainly used for liquids or melts but can also be implemented for some solid samples since the force is applied in a twisting motion. The instrument can do creep-recovery, stress-relaxation, and stress-strain experiments. Axial analyzers are used for solid or semisolid materials. It can do flexure, tensile, and compression testing (even shear and liquid specimens if desired). These analyzers can test higher modulus materials than torsional analyzers. The instrument can do thermomechanical analysis (TMA) studies in addition to the experiments that torsional analyzers can do.

Changing sample geometry and fixtures can make stress and strain analyzers virtually indifferent of one another except at the extreme ends of sample phases, i.e. really fluid or rigid materials. Common geometries and fixtures for axial analyzers include three-point and four-point bending, dual and single cantilever, parallel plate and variants, bulk, extension/tensile, and shear plates and sandwiches. Geometries and fixtures for torsional analyzers consist of parallel plates, cone-and-plate, couette, and torsional beam and braid. In order to utilize DMA to characterize materials, the fact that small dimensional changes can also lead to large inaccuracies in certain tests needs to be addressed. Inertia and shear heating can affect the results of either forced or free resonance analyzers, especially in fluid samples.

Evolved gas analysis

Evolved gas analysis (EGA) is a method used to study the gas evolved from a heated sample that undergoes decomposition or desorption. It is possible to detect which gas is evolved using evolved gas detection (EGD). EGD is often performed by coupling EGA with mass spectrometry, Fourier transform spectroscopy, gas chromatography, or Optical In-Situ Evolved Gas Analysis.

By coupling the thermal analysis instrument, e. g. TGA (Thermogravimetry) or DSC (Differential scanning calorimetry), with a fast Quadrupole Mass Spectrometer (QMS) the detection of gas separation and identification of the separated components are possible in exact time correlation with the other thermal analysis signals. DSC/TGA-QMS or TGA-QMS yields information on the composition (mass numbers of elements and molecules) of the evolved gases. It allows fast and easy interpretation of atomic/inorganic vapors and standard gases like H2, H2O, CO2, etc. Fragmentation, interpretation of organic molecules is sometimes difficult.

The combination with an FTIR (Fourier Transform Infrared Spectrometer) has become popular, especially in the polymer producing, chemical and pharmaceutical industry. DSC/TGA-FTIR or TGA-FTIR yields information on the composition (absorption bands) of the evolved gases (bonding conditions). The advantage is an easy interpretation (spectra data bases) of organic vapors without fragmentation. Symmetrical molecules can not be detected.

An EGA instrument named the Thermal and Evolved-Gas Analyzer was flown on the Phoenix Lander probe that reached Mars in May 2008. Its purpose was to study Martian soil samples.

Laser flash analysis

The laser flash analysis or laser flash method is used to measure thermal diffusivity of a multiplicity of different materials. An energy pulse heats one side of a plane-parallel sample. The temperature rise on the backside due to the energy input is time-dependent detected. The higher the thermal diffusivity of the sample, the faster the energy reaches the backside. A state-of-the-art laser flash apparatus (LFA) to measure thermal diffusivity over a broad temperature range, is shown on the right hand side.

Measurement principle

The laser flash method was developed by Parker et al. in 1961. In a vertical setup a light source (e.g. laser, flashlamp) heats the sample from the bottom side and a detector on top detects the time-depedent temperature rise. For measuring the thermal diffusivity, which is strongly temperature-dependent, at different temperatures the sample can be placed in a furnace at constant temperature.

Perfect conditions are

- Homogenous material,
- A homogenous energy input on the front side
- A time-dependent short pulse - in form of a Dirac delta function

Several improvements on the models have been made. In 1963 Cowan takes radiation and convection on the surface into account. Cape and Lehman consider transient heat transfer, finite pulse effects and also heat losses in the same year. Blumm and Opfermann improved the Cape-Lehman-Model with high order solutions of radial transient heat transfer and facial heat loss, non-linear regression routine in case of high heat losses and an advanced, patented pulse length correction.

Thermogravimetric Analysis

Thermogravimetric analysis or thermal gravimetric analysis (TGA) is a method of thermal analysis in which changes in physical and chemical properties of materials are measured as a function of increasing temperature (with constant heating rate), or as a function of time (with constant temperature and/or constant mass loss). TGA can provide information about physical phenomena, such as second-order phase transitions, including vaporization, sublimation, absorption, adsorption, and desorption. Likewise, TGA can provide information about chemical phenomena including chemisorptions, desolvation (especially dehydration), decomposition, and solid-gas reactions (e.g., oxidation or reduction).

TGA is commonly used to determine selected characteristics of materials that exhibit either mass loss or gain due to decomposition, oxidation, or loss of volatiles (such as moisture). Common applications of TGA are (1) materials characterization through analysis of characteristic decomposition patterns, (2) studies of degradation mechanisms and reaction kinetics, (3) determination of organic content in a sample, and (4) determination of inorganic (e.g. ash) content in a sample, which may be useful for corroborating predicted material structures or simply used as a chemical analysis. It is an especially useful technique for the study of polymeric materials, including thermoplastics, thermosets, elastomers, composites, plastic films, fibers, coatings and paints. Discussion of the TGA apparatus, methods, and trace analysis will be elaborated upon below. Thermal stability, oxidation, and combustion, all of which are possible interpretations of TGA traces, will also be discussed.

Instrumental Apparatus

Thermogravimetric analysis relies on a high degree of precision in three measurements: mass change, temperature, and temperature change. Therefore, the basic instrumental requirements for TGA are a precision balance with a pan loaded with the sample, and a programmable furnace. The furnace can be programmed either for a constant heating rate, or for heating to acquire a constant mass loss with time.

Though a constant heating rate is more common, a constant mass loss rate can illuminate specific reaction kinetics. For example, the kinetic parameters of the carbonization of polyvinyl butyral were found using a constant mass loss rate of 0.2 wt %/min. Regardless of the furnace programming, the sample is placed in a small, electrically heated furnace equipped with a thermocouple to monitor accurate measurements of the temperature by comparing its voltage output with that of the voltage-versus-temperature table stored in the computer's memory. A reference sample may be placed on another balance in a separate chamber. The atmosphere in the sample chamber may be purged with an inert gas to prevent oxidation or other undesired reactions. A different process using a quartz crystal microbalance has been devised for measuring smaller samples on the order of a microgram (versus milligram with conventional TGA). Figure 1 provides schematic diagrams of a typical TGA instrument.

Methods

The TGA instrument continuously weighs a sample as it is heated to temperatures of up to 2000°C for coupling with FTIR and Mass spectrometry gas analysis. As the temperature increases, various components of the sample are decomposed and the weight percentage of each resulting mass change can be measured. Results are plotted with temperature on the X-axis and mass loss on the Y-axis. The data can be adjusted using curve smoothing and first derivatives are often also plotted to determine points of inflection for more in-depth interpretations (see discussion on Trace Analysis).

Thermal Stability

TGA can be used to evaluate the thermal stability of a material. In a desired temperature range, if a species is thermally stable, there will be no observed mass change. Negligible mass loss corresponds to little or no slope in the TGA trace. TGA also gives the upper use temperature of a material. Beyond this temperature the material will begin to degrade.

TGA has a wide variety of applications, including analysis of ceramics and thermally stable polymers. Ceramics usually melt before they decompose as they are thermally stable over a large temperature range, thus TGA is mainly used to investigate the thermal stability of polymers. Most polymers melt or degrade before 200°C. However, there is a class of thermally stable polymers that are able to withstand temperatures of at least 300°C in air and 500°C in inert gases without structural changes or strength loss, which can be analyzed by TGA. For example, the polyimide Kapton loses less than 10% mass when held in 400°C air for 100 hours.

Compares various high performance fibers using TGA as an evaluation of thermal stability. From the TGA, polyoxazole (PBO) has the highest thermal stability of the four fibers as it is stable up to ca. 500°C. Ultra-high-molecular-weight polyethylene (UHMW-PE) has the lowest thermal stability, as it begins to degrade around 200°C. High performance fibers used in bulletproof vests must remain strong enough mechanically so as to protect the user from incoming projectiles. The thermal and photochemical degradation of the fibers causes the mechanical properties of the vests to decrease, effectively rendering the armor useless. Thus, thermal stability is a key property when designing these vests.

Three ways a material can lose mass during heating are through chemical reactions, the release of adsorbed species, and decomposition. All of these indicate that the material is no longer thermally stable. Because the TGA is performed in air, oxygen reacts with the organic fibers which eventually degrade completely, evidenced by the 100% mass loss. It is important to link thermal stability to the gas in which the TGA is performed. PBO, which completely decomposes when heated in air, retains ~60% mass when heated in N2. Thus, PBO is thermally stable in nitrogen up to 630°C, whereas in air, PBO has almost completely decomposed at that temperature.

Combustion during TG analysis is identifiable by distinct traces made in the TGA thermograms produced. One interesting example occurs with samples of as-produced unpurified carbon nanotubes that have a large amount of metal catalyst present. Due to combustion, a TGA trace can deviate from the normal form of a well-behaved function. This phenomenon arises from a rapid temperature change. When the weight and temperature are plotted versus time, a dramatic slope change in the first derivative plot is concurrent with the mass loss of the sample and the sudden increase in temperature seen by the thermocouple. The mass loss could be the result of particles of smoke released from burning caused by inconsistencies in the material itself, beyond the oxidation of carbon due to poorly controlled weight loss.

THERMOMECHANICAL ANALYSIS

Thermomechanical analysis (TMA) is a technique used in thermal analysis, a branch of materials science which studies the properties of materials as they change with temperature. Thermomechanical analysis is a subdiscipline of the themomechanometry (TM) technique.

Thermomechanometry is the measurement of a change of a dimension or a mechanical property of the sample while it is subjected to a temperature regime. An associated thermoanalytical method is thermomechanical analysis. A special related technique is thermodilatometry (TD), the measurement of a change of a dimension of the sample with a negligible force acting on the sample while it is subjected to a temperature regime. The associated thermoanalytical method is thermodilatometric analysis (TDA).

TDA is often referred to as zero force TMA. The temperature regime may be heating, cooling at a rate of temperature change that can include stepwise temperature changes, linear rate of change, temperature modulation with a set frequency and amplitude, free (uncontrolled) heating or cooling, or maintaining a constant increase in temperature. The sequence of temperatures with respect to time may be predetermined (temperature programmed) or sample controlled (controlled by a feedback signal from the sample response).

Thermomechanometry includes several variations according to the force and the way the force is applied.

Static force TM (sf-TM) is when the applied force is constant; previously called TMA with TD as the special case of zero force.

Dynamic force TM (df-TM) is when the force is changed as for the case of a typical stress–strain analysis; previously called TMA with the term dynamic meaning any alteration of the variable with time, and not to be confused with dynamic mechanical analysis (DMA).

Modulated force TM (mf-TM) is when the force is changed with a frequency and amplitude; previously called DMA. The term modulated is a special variant of dynamic, used to be consistent with modulated temperature differential scanning calorimetry (mt-DSC) and other situations when a variable is imposed in a cyclic manner.

Mechanical test

Mechanical testing seeks to measure mechanical properties of materials using various test specimen and fixture geometries using a range of probe types.

Measurement is desired to take place with minimal disturbance of the material being measured. Some characteristics of a material can be measured without disturbance, such as dimensions, mass, volume, density. However, measurement of mechanical properties normally involves disturbance of the system being measured.

The measurement often reflects the combined material and measuring device as the system. Knowledge of a structure can be gained by imposing an external stimulus and measuring the response of the material with a suitable probe. The external stimulus can be a stress or strain, however in thermal analysis the influence is often temperature.

Thermomechanometry is where a stress is applied to a material and the resulting strain is measured while the material is subjected to a controlled temperature program. The simplest mode of TM is where the imposed stress is zero. No mechanical stimulus is imposed upon the material, the material response is generated by a thermal stress, either by heating or cooling.

Zero force thermomechanometry

Zero force TM (a variant of sf-TM or TD) measures the response of the material to changes in temperature and the basic change is due to activation of atomic or molecular phonons. Increased thermal vibrations produce thermal expansion characterized by the coefficient of thermal expansion (CTE) that is the gradient of the graph of dimensional change versus temperature.

CTE depends upon thermal transitions such as the glass transition. CTE of the glassy state is low, while at the glass transition temperature (Tg) increased degrees of molecular segmental motion are released so CTE of the rubbery state is high. Changes in an amorphous polymer may involve other sub-Tg thermal transitions associated with short molecular segments, side-chains and branches. The linearity of the sf-TM curve will be changed by such transitions.

Other relaxations may be due to release of internal stress arising from the non-equilibrium state of the glassy amorphous polymer. Such stress is referred to as thermal aging. Other stresses may be as a result of moulding pressures, extrusion orientation, thermal gradients during solidification and externally imparted stresses.

Semi-crystalline polymers

Semi-crystalline polymers are more complex than amorphous polymers, since the crystalline regions are interspersed with amorphous regions. Amorphous regions in close association to the crystals or contain common molecules as tie molecules have less degrees of freedom than the bulk amorphous phase. These immobilised amorphous regions are called the rigid amorphous phase. CTE of the rigid amorphous phase is expected to be lower than that of the bulk amorphous phase.

The crystallite are typically not at equilibrium and they may contain different polymorphs. The crystals re-organize during heating so that they approach the equilibrium crystalline state. Crystal re-organization is a thermally activated process. Further crystallization of the amorphous phase may take place. Each of these processes with interfere with thermal expansion of the material.

The material may be a blend or a two-phase block or graft copolymer. If both phases are amorphous then two Tg will be observed if the material exists as two phases. If one Tg is exhibited then it will be between the Tg of the components and the resultant Tg will likely be described by a relationship such as the Flory-Fox or Kwei equations.

If one of the components is semi-crystalline then the complexity of a pure crystalline phase and either one or two amorphous phases will result. If both components are semi-crystalline then the morphology will be complex since both crystal phases will likely form separately, though with influence on each other.

Cross-linking

Cross-linking will restrict the molecular response to temperature change since degree of freedom for segmental motions are reduced as molecules become irreversibly linked. Cross-linking chemically links molecules, while crystallinity and fillers introduce physical constraints to motion. Mechanical properties such as derived from stress-strain testing are used to calculate crosslink density that is usually expressed as the molar mass between crosslinks (Mc).

The sensitivity of zero stress TMA to crosslinking is low since the structure receives minimum disturbance. Sensitivity to crosslinks requires high strain such that the segments between crosslinks become fully extended.

Zero force TM will only be sensitive to changes in the bulk that are expressed as a change in a linear dimension of the material. The measured change will be the resultant of all processes occurring as the temperature is changed. Some of the processes will be reversible, others irreversible, and others time-dependent. The methodology must be chosen to best detect, distinguish and resolve the thermal expansion or contractions observable.

The TM instrument need only apply sufficient stress to keep the probe in contact with the specimen surface, but it must have high sensitivity to dimensional change. The experiment must be conducted at a temperature change rate slow enough for the material to approach thermal equilibrium throughout. While the temperature should be the same throughout the material it will not necessarily be at thermal equilibrium in the context of molecular relaxations.

The temperature of the molecules relative to equilibrium is expressed as the fictive temperature. The fictive temperature is the temperature at which the unrelaxed molecules would be at equilibrium.

Zero-stress thermomechanometry experimental

TM is sufficient for zero stress experiments since superimposition of a frequency to create a dynamic mechanical experiment will have no effect since there is no stress other than a nominal contact stress. The material can be best characterized by an experiment in which the original material is first heated to the upper temperature required, then the material should be cooled at the same rate, followed by a second heating scan.

The first heating scan provides a measure of the material with all of its structural complexities. The cooling scan allows and measures the material as the molecules lose mobility, so it is going from an equilibrium state and gradually moving away from equilibrium as the cooling rate exceeds the relaxation rate. The second heating scan will differ from the first heating scan because of thermal relaxation during the first scan and the equilibration achieved during the cooling scan. A second cooling scan followed by a third heating scan can be performed to check on the reliability of the prior scans. Different heating and cooling rates can be used to produce different equilibrations. Annealing at specific temperatures can be used to provide different isothermal relaxations that can be measured by a subsequent heating scan.

Static-force TM

The sf-TM experiments duplicate experiments that can be performed using differential scanning calorimetry (DSC). A limitation of DSC is that the heat exchange during a process or due to the heat capacity of the material cannot be measured over long times or at slow heating or cooling rates since the finite quantity of heat exchanges will be dispersed over too long a time to be detected. The limitation does not apply to sf-TM since the dimensional change of the material can be measured over any time. The constraint is the practical time for the experiment. The application of multiple scans is shown above to distinguish reversible from irreversible changes. Thermal cycling and annealing steps can be added to provide complex thermal programs to test various attributes of a material as more becomes known about the material.

Modulated-temperature TM

Modulated temperature TM (mt-TM) has been used as an analogous experiment to modulated-temperature DSC (mtDSC). The principle of mt-TM is similar to the DSC analogy. The temperature is modulated as the TM experiment proceeds. Some thermal processes are reversible, such as the true CTE, while others such as stress relief, orientation randomization and crystallization are irreversible within the conditions of the experiment. The modulation conditions should be different from mt-DSC since the sample and test fixture and enclosure is larger thus requiring longer equilibration time. mt-DSC typically uses a period of 60 s, amplitude 0.5-1.0 °C and average heating or cooling rate of 2 °C·min-1. MT-TMA may have a period of 1000 s with the other parameters similar to mt-DSC. These conditions will require long scan times.

Another experiment is an isothermal equilibration where the material is heated rapidly to a temperature where relaxations can proceed more rapidly. Thermal aging can take several hours or more under ideal conditions. Internal stresses may relax rapidly. TM can be used to measure the relaxation rates and hence characteristic times for these events, provides they are within practical measurements times available for the instrument. Temperature is the variable that can be changed to bring relaxations into measurable time ranges.

STATIC FORCE THERMOMECHANOMETRY EXPERIMENTAL

Creep and stress relaxation measures the elasticity, viscoelasticity and viscous behaviour of materials under a selected stress and temperature. Tensile geometry is the most common for creep measurements. A small force is initially imparted to keep the specimen aligned and straight. The selected stress is applied rapidly and held constant for the required time; this may be 1 h or more. During application of force the elastic property is observed as an immediate elongation or strain. During the constant force period the time dependent elastic response or viscoelasticity, together with the viscous response, result in further increase in strain.

The force is removed rapidly, though the small alignment force is maintained. The recovery measurement time should be four times the creep time, so in this example the recovery time should be 4 h. Upon removal of the force the elastic component results in an immediate contraction. The viscoelastic recovery is exponential as the material slowly recovers some of the previously imparted creep strain. After recovery there is a permanent unrecovered strain due to the viscous component of the properties.

Analysis of the data is performed using the four component viscoelastic model where the elements are represented by combinations of springs and dashpots. The experiment can be repeated using different creep forces. The results for varying forces after the same creep time can be used to construct isochronal stress–strain curves. The creep and recovery experiment can be repeated under different temperatures. The creep–time curves measured at various temperatures can be extended using the time-temperature-superposition principle to construct a creep and recovery mastercurve that extends the data to very long and very short times. These times would be impractical to measure directly. Creep at very long timeframes is important for prediction of long term properties and product lifetimes. A complementary property is stress relaxation, where a strain is applied and the corresponding stress change is measured. The mode of measurement is not directly available with most thermomechanical instruments. Stress relaxation is available using any standard universal test instruments, since their mode of operation is application of strain, while the stress is measured.

Dynamic force thermomechanometry experimental

Experiments where the force is changed with time are called dynamic force thermomechanometry (df-TM). This use of the term dynamic is distinct from the situation where the force is periodically changed with time, typically following a sine relationship, where the term modulated is recommended. Most thermomechanical instruments are force controlled, that is they apply a force, then measure a resulting change in a dimension of the test specimen. Usually a constant strain rate is used for stress–strain measurements, but in the case of df-TM the stress will be applied at a chosen rate.

The result of a stress-strain analysis is a curve that will reveal the modulus (hardness) or compliance (softness, the reciprocal of modulus). The modulus is the slope of the initial linear region of the stress–strain curve. Various ways of selecting the region to calculate gradient are used such as the initial part of the curve, another is to selected a region defined by the secant to the curve. If the test material is a thermoplastic a yield zone may be observed and a yield stress (strength) calculated. A brittle material will break before it yields. A ductile material will further deform after yielding. When the material breaks a break stress (ultimate stress) and break strain are calculated. The area under the stress–strain curve is the energy required to break (toughness).

Thermomechanical instruments are distinct in that they can measure only small changes in linear dimension (typically 1 to 10 mm) so it is possible to measure yield and break properties for small specimens and those that do not change dimensions very much before exhibiting these properties.

A purpose of measuring a stress–strain curve is to establish the linear viscoelastic region (LVR). LVR is this initial linear part of a stress–strain curve where an increase in stress is accompanied by a proportional increase in strain, that is the modulus is constant and the change in dimension is reversible. A knowledge of LVR is a prerequisite for any modulated force thermomechanometry experiments. Conduct of complex experiments should be preceded by preliminary experiments with a limited range of variables to establish the behaviour of the test material for selection of further instrument configuration and operating parameters.

Modulated temperature thermomechanometry experimental

Modulated temperature conditions are where the temperature is changed in a cyclic manner such as in a sine, isothermal-heating, isothermal-cooling or heat-cool. The underlying temperature can increase, decrease or be constant. Modulated temperature conditions enable separation of the data into reversing data that is in-phase with the temperature changes, and non-reversing that is out-of-phase with the temperature changes. Sf-TM is required since the force should be constant while the temperature is modulated, or at least constant for each modulation period.

A reversing properties is coefficient of thermal expansion. Non-reversing properties are thermal relaxations, stress relief and morphological changes that occur during heating, causing the material to approach thermal equilibrium.

8

INTRODUCTION TO POLYMERS AND POLYMERIZATION

Polymers are a large class of materials consisting of many small molecules (called monomers) that can be linked together to form long chains, thus they are known as macromolecules. The picture at the top of the page is a short section of such a chain. A typical polymer may include tens of thousands of monomers. Because of their large size, polymers are classified as macromolecules.

Humans have taken advantage of the versitility of polymers for centuries in the form of oils, tars, resins, and gums. However, it was not until the industrial revolution that the modern polymer industry began to develop. In the late 1830s, Charles Goodyear succeeded in producing a useful form of natural rubber through a process known as "vulcanization." Some 40 years later, Celluloid (a hard plastic formed from nitrocellulose) was successfully commercialized. Despite these advances, progress in polymer science was slow until the 1930s, when materials such as vinyl, neoprene, polystyrene, and nylon were developed. The introduction of these revolutionary materials began an explosion in polymer research that is still going on today.

Unmatched in the diversity of their properties, polymers such as cotton, wool, rubber, Teflon(tm), and all plastics are used in nearly every industry. Natural and synthetic polymers can be produced with a wide range of stiffness, strength, heat resistance, density, and even price. With continued research into the science and applications of polymers, they are playing an ever increasing role in society. The following sections provide an introduction to the science of macromolecules.

A polymer is a chemical compound or mixture of compounds consisting of repeating structural units created through a process of polymerization. The term derives from the ancient Greek word ðïëýò (polus, meaning "many, much") and ìÝñïò (meros, meaning "parts"), and refers to a molecule whose structure is composed of multiple repeating units, from which originates a characteristic of high relative molecular mass and attendant properties. The units composing polymers derive, actually or conceptually, from molecules of low relative molecular mass. The term was coined in 1833 by Jöns Jacob Berzelius, though with a definition distinct from the modern IUPAC definition. Polymers are studied in the fields of biophysics and macromolecular science, and polymer science (which includes polymer chemistry and polymer physics).

Historically, products arising from the linkage of repeating units by covalent chemical bonds have been the primary focus of polymer science; emerging important areas of the science now focus on non-covalent links. Because of the stipulation as to repeating substructures, polymers are formally a subclass of the category of macromolecules; the polyisoprene of latex rubber and the polystyrene of styrofoam are examples of polymeric natural/biological and synthetic polymers, respectively. In biological contexts, essentially all biological macromolecules—i.e., proteins (polyamides), nucleic acids (polynucleotides), and polysaccharides—are purely polymeric, or are composed in large part of polymeric components—e.g., isoprenylated/lipid-modified glycoproteins, where small lipidic molecule and oligosaccharide modifications occur on the polyamide backbone of the protein.

Hence, the terms polymer and polymeric material encompass very large, broad classes of compounds, both natural and synthetic, with a wide variety of properties. Because of the extraordinary range of properties of polymeric materials, they play an essential and ubiquitous roles in everyday life, from those of familiar synthetic plastics and other materials of day-to-day work and home life, to the natural biopolymers that are fundamental to biological structure and function.

STEP-GROWTH

Step-growth polymers are defined as polymers formed by the stepwise reaction between functional groups of monomers, usually containing heteroatoms such as nitrogen or oxygen. Most step-growth polymers are also classified as condensation polymers, but not all step-growth polymers (like polyurethanes formed from isocyanate and alcohol bifunctional monomers) release condensates; in this case, we talk about addition polymers. Step-growth polymers increase in molecular weight at a very slow rate at lower conversions and reach moderately high molecular weights only at very high conversion (i.e., >95%).

Step-growth polymerization refers to a type of polymerization mechanism in which bi-functional or multifunctional monomers react to form first dimers, then trimers, longer oligomers and eventually long chain polymers. Many naturally occurring and some synthetic polymers are produced by step-growth polymerization, e.g. polyesters, polyamides, polyurethanes, etc. Due to the nature of the polymerization mechanism, a high extent of reaction is required to achieve high molecular weight. The easiest way to visualize the mechanism of a step-growth polymerization is a group of people reaching out to hold their hands to form a human chain — each person has two hands (= reactive sites). There also is the possibility to have more than two reactive sites on a monomer: In this case branched polymers are produced.

To alleviate inconsistencies in these naming methods, adjusted definitions for condensation and addition polymers have been developed. A condensation polymer is defined as a polymer that involves loss of small molecules during its synthesis, or contains heteroatoms as part of its backbone chain, or its repeat unit does not contain all the atoms present in the hypothetical monomer to which it can be degraded.

Chain-growth

Chain-growth polymerization (or addition polymerization) involves the linking together of molecules incorporating double or triple carbon-carbon bonds. These unsaturated monomers (the identical molecules that make up the polymers) have extra internal bonds that are able to break and link up with other monomers to form a repeating chain, whose backbone typically contains only carbon atoms. Chain-growth polymerization is involved in the manufacture of polymers such as polyethylene, polypropylene, and polyvinyl chloride (PVC). A special case of chain-growth polymerization leads to living polymerization.

In the radical polymerization of ethylene, its ð bond is broken, and the two electrons rearrange to create a new propagating center like the one that attacked it. The form this propagating center takes depends on the specific type of addition mechanism. There are several mechanisms through which this can be initiated. The free radical mechanism is one of the first methods to be used. Free radicals are very reactive atoms or molecules that have unpaired electrons. Taking the polymerization of ethylene as an example, the free radical mechanism can be divided in to three stages: chain initiation, chain propagation, and chain termination.

Chain growth polymerization is a polymerization technique where unsaturated monomer molecules add onto the active site on a growing polymer chain one at a time. There are a limited number of these active sites at any moment during the polymerization which gives this method its key characteristics.

Free radical addition polymerization of ethylene must take place at high temperatures and pressures, approximately 300 °C and 2000 atm. While most other free radical polymerizations do not require such extreme temperatures and pressures, they do tend to lack control. One effect of this lack of control is a high degree of branching. Also, as termination occurs randomly, when two chains collide, it is impossible to control the length of individual chains. A newer method of polymerization similar to free radical, but allowing more control involves the Ziegler-Natta catalyst, especially with respect to polymer branching.

It can be represented with the chemical equation:

where n is the degree of polymerization and M is some form of unsaturated compound: an Alkene (vinyl polymers) or Alicyclic compound (ring-opening polymerization) containing molecule.

Chain-growth polymerization usually has the following steps:

1. Chain initiation, usually by means of an initiator which starts the chemical process. Typical initiators include any organic compound with a labile group: e.g. azo (-N=N-), disulfide (-S-S-), or peroxide (-O-O-). Two examples are benzoyl peroxide and AIBN.
2. Chain propagation
3. Chain transfer, terminates the chain, but the active site is transferred to a new chain. This can occur with the solvent, monomer, or other polymer. This process increases the branching of the resulting polymer.
4. Chain termination, which occurs either by combination or disproportionation. Termination, in radical polymerization, is when the free radicals combine and is the end of the polymerization process.

The active center can be one of a number of different types:

- Free radical in radical polymerization, for example, polystyrene, sometimes seen as packing peanuts, is produced by polymerizing styrene with Benzoyl peroxide as its radical initiator
- Carbocation in cationic polymerization, an example is Isobutyl synthetic rubber, initiated by Aluminium chloride ionizing isobutylene
- Carbanion in anionic polymerization
- Organometallic complex in coordination polymerization.

Under the necessary reaction conditions, an addition polymerization can be considered a living polymerization. This is most often seen with anionic polymerization as it can be easy to perform without termination steps.

Chain-growth polymerization result in high molecular weight polymer being formed at low conversion. This final weight is determined by the rate of propagation compared to the rate of individual chain termination, which includes both chain transfer and chain termination steps. above a certain ceiling temperature, no polymerization occurs.

Other forms of chain growth polymerization include cationic addition polymerization and anionic addition polymerization. While not used to a large extent in industry yet due to stringent reaction conditions such as lack of water and oxygen, these methods provide ways to polymerize some monomers that cannot be polymerized by free radical methods such as polypropylene. Cationic and anionic mechanisms are also more ideally suited for living polymerizations, although free radical living polymerizations have also been developed.

Esters of acrylic acid contain a carbon-carbon double bond which is conjugated to an ester group. This allows the possibility of both types of polymerization mechanism. An acrylic ester by itself can undergo chain-growth polymerization to form a homopolymer with a carbon-carbon backbone, such as poly(methyl methacrylate). Also, however, certain acrylic esters can react with diamine monomers by nucleophilic conjugate addition of amine groups to acrylic C=C bonds. In this case the polymerization proceeds by step-growth and the products are poly(beta-amino ester) copolymers, with backbones containing nitrogen (as amine) and oxygen (as ester) as well as carbon.

COMPARISON WITH OTHER POLYMERIZATION METHODS

The distinction between step-growth polymerization and chain-growth polymerization was instead introduced by Paul Flory in 1953, and refers to the difference in reaction mechanisms with step-growth using the functional groups of the monomer compared to the free-radical or ion groups used in chain-growth polymerization.

Chain growth polymerization and addition polymerization (also called polyaddition) are two different concepts. In fact polyurethane polymerizes with addition polymerization (because its polymerization does not produce any small molecules, called "condensate"), but its reaction mechanism is a step-growth polymerization.

The distinction between "addition polymerization" and "condensation polymerization" was introduced by Wallace Hume Carothers in 1929, and refers to the type of product produced. Addition polymerization produces only a polymer molecule, while condensation polymerization produces a polymer as well as a molecule with a low molecular weight, usually water.

Plasma polymerization

Plasma polymerization (or glow discharge polymerization) uses plasma sources to generate a gas discharge that provides energy to activate or fragment gaseous or liquid monomer, often containing a vinyl group, in order to initiate polymerization. Polymers formed from this technique are generally highly branched and highly cross-linked, and adhere to solid surfaces well. The biggest advantage to this process is that polymers can be directly attached to a desired surface while the chains are growing, which reduces steps necessary for other coating processes such as grafting. This is very useful for pinhole-free coatings of 100 picometers to 1 micrometre thickness with solvent insoluble polymers.

In as early as the 1870s "polymers" formed by this process were known, but these polymers were initially thought of as undesirable byproducts associated with electric discharge, with little attention being given to their properties. It was not until the 1960s that the properties of these polymers where found to be useful. It was found that flawless thin polymeric coatings could be formed on metals. By selecting the monomer type and the energy density per monomer, known as the Yasuda parameter, the chemical composition and structure of the resulting thin film can be varied with a wide range. These films are usually inert, adhesive, and have low dielectric constants. Some common monomers polymerized by this method include styrene, ethylene, methacrylate and pyridine, just to name a few. The 1970s brought about many advances in plasma polymerization, including the polymerization of many different types of monomers, and a probable mechanism of the polymerization process being determined. Since this time most attention devoted to plasma polymerization has been in the fields of coatings, but since it is difficult to control polymer structure, it has very limited applications.

Types of reactors

There are a few designs for apparatus used in plasma polymerization, one of which is the Bell (static type), in which monomer gas is put into the reaction chamber, but does not flow through the chamber. It comes in and polymerizes without removal. This reactor has internal electrodes, and polymerization generally takes place on the cathode side. All devices contain the thermostatic bath, which is used to regulate temperature, and a vacuum to regulate pressure.

Operation: The monomer gas comes into the Bell type reactor as a gaseous species, and then is put into the plasma state by the electrodes, in which the plasma may consist of radicals, anions and cations. These monomers are then polymerized on the cathode surface, or some other surface placed in the apparatus by different mechanisms of which details are discussed below. The deposited polymers then propagate off the surface and form growing chains with seemingly uniform consistency.

Another popular reactor type is the flow through reactor (continuous flow reactor), which also has internal electrodes, but this reactor allows monomer gas to flow through the reaction chamber as its name implies, which should give a more even coating for polymer film deposition. It has the advantage that more monomer keeps flowing into the reactor in order to deposit more polymer. It has the disadvantage of forming what is called "tail flame," which is when polymerization extends into the vacuum line.

A third popular type of reactor is the electrodeless. This uses an RF coil wrapped around the glass apparatus, which then uses a radio frequency generator to form the plasma inside of the housing without the use of direct electrodes. The polymer can then be deposited as it is pushed through this RF coil toward the vacuum end of the apparatus. This has the advantage of not having polymer building up on the electrode surface, which is desirable when polymerizing onto other surfaces.

A fourth type of system growing in popularity is the atmospheric-pressure plasma system, which is useful for depositing thin polymer films. This system bypasses the requirements for special hardware involving vacuums, which then makes it favorable for integrated industrial use. It has been shown that that polymers formed at atmospheric-pressure can have similar properties for coatings as those found in the low-pressure systems.

PHYSICAL PROCESS CHARACTERISTICS

The formation of a plasma for polymerization depends on many of the following. An electron energy of 1–10 eV is required, with electron densities of 109 to 1012 per cubic centimeter, in order to form the desired plasma state. The formation of a low-temperature plasma is important; the electron temperatures are not equal to the gas temperatures and have a ratio of Te/Tg of 10 to 100, so that this process can occur at near ambient temperatures, which is advantageous because polymers degrade at high temperatures, so if a high-temperature plasma was used the polymers would degrade after formation or would never be formed. This entails non-equilibrium plasmas, which means that charged monomer species have more kinetic energy than neutral monomer species, and cause the transfer of energy to a substrate instead of an uncharged monomer.

Kinetics

The kinetic rate of these reactions depends mostly on the monomer gas, which must be either gaseous or vaporized. However, other parameters are also important as well, such as power, pressure, flow rate, frequency, electrode gap and reactor configuration. Low flow rates usually only depend on the amount of reactive species present for polymerization, whereas high flow rates depend on the amount of time that is spent in the reactor. Therefore, the maximum rate of polymerization is somewhere in the middle.

The fastest reactions tend to be in the order of triple-bonded > double-bonded > single bonded molecules, and also lower molecular weight molecules are faster than higher ones. So acetylene is faster than ethylene, and ethylene is faster than propene, etc. The molecular weight factor in polymer deposition is dependent on the monomer flow rate, in which a higher molecular weight monomer typically near 200 g/mol needs a much higher flow rate of 15×104 g/cm2, whereas lower molecular weights around 50 g/mol require a flow rate of only 5×104 g/cm2. A heavy monomer therefore needs a faster flow, and would likely lead to increased pressures, decreasing polymerization rates.

Increased pressure tends to decrease polymerization rates reducing uniformity of deposition since uniformity is controlled by constant pressure. This is a reason that high-pressure plasma or atmospheric-pressure plasmas are not usually used in favor of low-pressure systems, however atmospheric-pressure systems are still becoming more popular due to their ease of incorporation in manufacturing systems, with similar results now being formed compared to low pressure systems. At pressures greater than 1 torr, oligomers are formed on the electrode surface, and the monomers also on the surface can dissolve them to get a low degree of polymerization forming an oily substance. At low pressures, the reactive surfaces are low in monomer and facilitate growing high molecular weight polymers.

The rate of polymerization depends on input power, until power saturation occurs and the rate becomes independent of it. A narrower electrode gap also tends to increase polymerization rates because a higher electron density per unit area is formed. Polymerization rates also depend on the type of apparatus used for the process. In general, increasing the frequency of alternating current glow discharge up to about 5 kHz increases the rate due to the formation of more free radicals. After this frequency, inertial effects of colliding monomers inhibit polymerization.

This forms the first plateau for polymerization frequencies. A second maximum in frequency occurs at 6 MHz, where side reactions are overcome again and the reaction occurs through free radicals diffused from plasma to the electrodes, at which point a second plateau is obtained. These parameters differ slightly for each monomer and must be optimized in-situ.

Plasma contains many species such as ions, free radicals and electrons, so it is important to look at what contributes to the polymerization process most. The first suggested process by Westwood et al. was that of a cationic polymerization, since in a direct current system polymerization occurs mainly on the cathode. However, more investigation has led to the belief that the mechanism is more of a radical polymerization process, since radicals tend to be trapped in the films, and termination can be overcome by reinitiation of oligomers. Other kinetic studies also support this theory.

In polymerization, both gas phase and surface reactions occur, but mechanism differs between high and low frequencies. At high frequencies it occurs in radical intermediates, whereas at low frequencies polymerization happens mainly on surfaces. As polymerization occurs, the pressure inside the chamber decreases in a closed system, since gas phase monomers go to solid polymers.

Polymers can be placed on many substrates other than the electrode surfaces, such as glass, other organic polymers or metals, when either a surface is placed in front of the electrodes, or placed in the middle between them. The ability for them to build off of electrode surfaces is likely to be an electrostatic interaction, while on other surfaces covalent attachment is possible.

The actual polymerization is likely to take place through many ionic and radical processes which are initiated by plasma formed from the glow discharge.

GENERAL CHARACTERISTICS OF PLASMA POLYMERS

The properties of plasma polymers differ greatly from those of conventional polymers. While both types are dependent on the chemical properties of the monomer, the properties of plasma polymers depend more greatly on the design of the reactor and the chemical and physical characteristics of the substrate on which the plasma polymer is deposited. The location within the reactor where the deposition occurs also has an effect on the resultant polymer's properties. In fact by using plasma polymerization with a single monomer and varying the reactor, substrate, etc. a variety of polymers, each having different physical and chemical properties, can be prepared. The large dependence of the polymer features on these factors make it difficult to assign a set of basic characteristics, but a few common properties that set plasma polymers apart from conventional polymers do exist.

The most significant difference between conventional polymers and plasma polymers is that plasma polymers do not contain regular repeating units. Due to the number of different propagating species present at any one time as discussed above, the resultant polymer chains are highly-branched and are randomly terminated with a high degree of cross-linking. An example of a proposed structure for plasma polymerized ethylene demonstrating a large extend of cross-linking and branching is shown in Figure 4.

All plasma polymers contain free radicals as well. The amount of free radicals present varies between polymers and is dependent on the chemical structure of the monomer. Because the formation of the trapped free radicals is tied to the growth mechanism of the plasma polymers, the overall properties of the polymers directly correlate to the number of free radicals. Plasma polymers also contain an internal stress. If a thick layer (e.g. 1 μm) of a plasma polymer is deposited on a glass slide, the plasma polymer will buckle and frequently crack. The curling is attributed to an internal stress formed in the plasma polymer during the polymer deposition. The degree of curling is dependent on the monomer as well as the conditions of the plasma polymerization.

Most plasma polymers are insoluble and infusible. These properties are due to the large amount of cross-linking in the polymers, previously discussed. Consequently the kinetic path length for these polymers must be sufficiently long, so these properties can be controlled to a point.

The permeabilities of plasma polymers also differ greatly from those of conventional polymers. Because of the absence of large-scale segmental mobility and the high degree of cross-linking within the polymers, the permeation of small molecules does not strictly follow the typical mechanisms of "solution-diffusion" or molecular-level sieve for such small permeants. Really the permeability characteristics of plasma polymers falls between these two ideal cases.

A final common characteristic of plasma polymers is the adhesion ability. The specifics of the adhesion ability for a given plasma polymer, such as thickness and characteristics of the surface layer, again are particular for a given plasma polymer and few generalizations can be made.

Advantages and disadvantages

Plasma polymerization offers a number of advantages over other polymerization methods and in general. The most significant advantage of plasma polymerization is its ability to produce polymer films of organic compounds that do not polymerize under normal chemical polymerization conditions. Nearly all monomers, even saturated hydrocarbons and organic compounds without a polymerizable structure such as a double bond, can be polymerized with this technique.

A second advantage is the ease of application of the polymers as coatings versus conventional coating processes. While coating a substrate with conventional polymers requires a number of steps, plasma polymerization accomplishes all these in essentially a single step. This leads to a cleaner and 'greener' synthesis and coating process, since no solvent is needed during the polymer preparation and no cleaning of the resultant polymer is needed either. Another 'green' aspect of the synthesis is that no initiator is needed for the polymer preparation since reusable electrodes cause the reaction to proceed. The resultant polymer coatings also have a number of advantages over typical coatings. These advantages include being nearly pinhole free, highly dense, and that the thickness of the coating can easily be varied.

There are also a number of disadvantages relating to plasma polymerization versus conventional methods. The most significant disadvantage is the high cost of the process. A vacuum system is required for the polymerization, significantly increasing the set up price.

Another disadvantage is due to the complexity of plasma processes. Because of the complexity it is not easy to achieve a good control over the chemical composition of the surface after modification. The influence of process parameters on the chemical composition of the resultant polymer means it can take a long time to determine the optimal conditions. The complexity of the process also makes it impossible to theorize what the resultant polymer will look like, unlike conventional polymers which can be easily determined based on the monomer.

9

LAWS OF THERMODYNAMICS

To be specific, it explains how thermal energy is converted to or from other forms of energy and how matter is affected by this process. Thermal energy is the energy that comes from heat. This heat is generated by the movement of tiny particles within an object, and the faster these particles move, the more heat is generated.

Thermodynamics is the study of the energy, principally heat energy, that accompanies chemical or physical changes. Some chemical reactions release heat energy; they are called exothermic reactions, and they have a negative enthalpy change. Others absorb heat energy and are called endothermic reactions, and they have a positive enthalpy change. But thermodynamics is concerned with more than just heat energy. The change in level of organization or disorganization of reactants and products as changes take place is described by the entropy change of the process. For example, the conversion of one gram of liquid water to gaseous water is in the direction of increasing disorder, the molecules being much more disorganized as a gas than as a liquid. The increase in disorder is described as an increase in entropy, and the change in entropy is positive.

Whether a chemical reaction or physical change will occur depends on both the enthalpy and entropy of the process, which are quantities that can be calculated from tabulated data. Both terms are combined in the free energy—the third and most important thermodynamic term. If the change in free energy is negative, the reaction will proceed to the right; this reaction is called a spontaneous reaction. If the sign is positive, the reaction will not proceed as written; this reaction is nonspontaneous. A powerful prediction as to whether a reaction will or will not take place can be made using tabulated data to calculate the change in free energy.

Chemical thermodynamics is the study of the interrelation of heat and work with chemical reactions or with physical changes of state within the confines of the laws of thermodynamics. Chemical thermodynamics involves not only laboratory measurements of various thermodynamic properties, but also the application of mathematical methods to the study of chemical questions and the spontaneity of processes.

The structure of chemical thermodynamics is based on the first two laws of thermodynamics. Starting from the first and second laws of thermodynamics, four equations called the "fundamental equations of Gibbs" can be derived. From these four, a multitude of equations, relating the thermodynamic properties of the thermodynamic system can be derived using relatively simple mathematics. This outlines the mathematical framework of chemical thermodynamics.

Thermodynamics, science of the relationship between heat, work, temperature, and energy. In broad terms, thermodynamics deals with the transfer of energy from one place to another and from one form to another. The key concept is that heat is a form of energy corresponding to a definite amount of mechanical work.

Heat was not formally recognized as a form of energy until about 1798, when Count Rumford (Sir Benjamin Thompson), a British military engineer, noticed that limitless amounts of heat could be generated in the boring of cannon barrels and that the amount of heat generated is proportional to the work done in turning a blunt boring tool. Rumford's observation of the proportionality between heat generated and work done lies at the foundation of thermodynamics. Another pioneer was the French military engineer Sadi Carnot, who introduced the concept of the heat-engine cycle and the principle of reversibility in 1824. Carnot's work concerned the limitations on the maximum amount of work that can be obtained from a steam engine operating with a high-temperature heat transfer as its driving force. Later that century, these ideas were developed by Rudolf Clausius, a German mathematician and physicist, into the first and second laws of thermodynamics, respectively.

Different Branches of Thermodynamics

Thermodynamics is classified into the following four branches:

Classical Thermodynamics

Statistical Thermodynamics

Chemical Thermodynamics

Equilibrium Thermodynamics

CLASSICAL THERMODYNAMICS

In classical thermodynamics, the behaviour of matter is analysed with a macroscopic approach. Units such as temperature and pressure are taken into consideration, which helps the individuals calculate other properties and predict the characteristics of the matter undergoing the process.

Statistical Thermodynamics

In statistical thermodynamics, every molecule is under the spotlight, i.e. the properties of every molecule and how they interact are taken into consideration to characterise the behaviour of a group of molecules.

Chemical Thermodynamics

Chemical thermodynamics is the study of how work and heat relate to each other in chemical reactions and in changes of states.

Equilibrium Thermodynamics

Equilibrium thermodynamics is the study of transformations of energy and matter as they approach the state of equilibrium.

Thermodynamics: laws

Energy exists in many forms, such as heat, light, chemical energy, and electrical energy. Energy is the ability to bring about change or to do work. Thermodynamics is the study of energy.

The four laws of thermodynamics define fundamental physical quantities (temperature, energy, and entropy) that characterize thermodynamic systems. The laws describe how these quantities behave under various circumstances, and forbid certain phenomena (such as perpetual motion).

The four laws of thermodynamics are:

- Zeroth law of thermodynamics: If two systems are both in thermal equilibrium with a third system then they are in thermal equilibrium with each other. This law helps define the notion of temperature.
- First law of thermodynamics: Heat and work are forms of energy transfer. Energy is invariably conserved however the internal energy of a closed system may change as heat is transferred into or out of the system or work is done on or by the system. In real systems work does not always leave the system. For example, changes in molecular energy (potential energy), are generally considered to remain within the system. Similarly, the rotational and vibrational energies of polyatomic molecules remain

within the system. Energy can be changed from one form to another, but it cannot be created or destroyed. The total amount of energy and matter in the Universe remains constant, merely changing from one form to another. The First Law of Thermodynamics (Conservation) states that energy is always conserved, it cannot be created or destroyed. In essence, energy can be converted from one form into another.

From the above, all the energy associated with a system must be accounted for as heat, work, chemical energy etc., thus perpetual motion machines of the first kind, which would do work without using the energy resources of a system, are impossible.

- Second law of thermodynamics: An isolated system, if not already in its state of thermodynamic equilibrium, spontaneously evolves towards it. Thermodynamic equilibrium has the greatest entropy amongst the states accessible to the system. Perpetual motion machines of the second kind are thus impossible.

The Second Law of Thermodynamics states that "in all energy exchanges, if no energy enters or leaves the system, the potential energy of the state will always be less than that of the initial state." This is also commonly referred to as entropy. A watchspring-driven watch will run until the potential energy in the spring is converted, and not again until energy is reapplied to the spring to rewind it. A car that has run out of gas will not run again until you walk 10 miles to a gas station and refuel the car. Once the potential energy locked in carbohydrates is converted into kinetic energy (energy in use or motion), the organism will get no more until energy is input again. In the process of energy transfer, some energy will dissipate as heat. Entropy is a measure of disorder: cells are NOT disordered and so have low entropy. The flow of energy maintains order and life. Entropy wins when organisms cease to take in energy and die.

- Third law of thermodynamics: The entropy of a system approaches a constant value as the temperature approaches zero. The entropy of a system at absolute zero is typically zero, and in all cases is determined only by the number of different ground states it has. Specifically, the entropy of a pure crystalline substance at absolute zero temperature is zero.

Classical thermodynamics describes the exchange of work and heat between closed systems. It has a special interest in systems that are individually in states of thermodynamic equilibrium. Thermodynamic equilibrium is a condition of systems which are adequately described by only macroscopic variables. Every physical system, however, when microscopically examined, shows apparently random microscopic statistical fluctuations in its thermodynamic variables of state (entropy, temperature, pressure, etc.). These microscopic fluctuations are negligible for systems which are nearly in thermodynamic equilibrium and which are only macroscopically examined. They become important, however, for systems which are nearly in thermodynamic equilibrium when they are microscopically examined, and, exceptionally, for macroscopically examined systems that are in critical states, and for macroscopically examined systems that are far from thermodynamic equilibrium.

There have been suggestions of additional laws, but none of them achieve the generality of the four accepted laws, and they are not mentioned in standard textbooks.

The laws of thermodynamics are important fundamental laws in physics and they are applicable in other natural sciences.

Zeroth law

The zeroth law of thermodynamics may be stated in the following form:

If two systems are both in thermal equilibrium with a third then they are in thermal equilibrium with each other.

The law is intended to allow the existence of an empirical parameter, the temperature, as a property of a system such that systems in thermal equilibrium with each other have the same temperature. The law as stated here is compatible with the use of a particular physical body, for example a mass of gas, to match temperatures of other bodies, but does not justify regarding temperature as a quantity that can be measured on a scale of real numbers.

Though this version of the law is one of the more commonly stated, it is only one of a diversity of statements that are labeled as “the zeroth law” by competent writers. Some statements go further so as to supply the important physical fact that temperature is one-dimensional, that one can conceptually arrange bodies in real number sequence from colder to hotter. Perhaps there exists no unique “best possible statement” of the “zeroth law”, because there is in the literature a range of formulations of the principles of thermodynamics, each of which call for their respectively appropriate versions of the law.

Although these concepts of temperature and of thermal equilibrium are fundamental to thermodynamics and were clearly stated in the nineteenth century, the desire to explicitly number the above law was not widely felt until Fowler and Guggenheim did so in the 1930s, long after the first, second, and third law were already widely understood and recognized. Hence it was numbered the zeroth law. The importance of the law as a foundation to the earlier laws is that it allows the definition of temperature in a non-circular way without reference to entropy, its conjugate variable. Such a temperature definition is said to be 'empirical'.

First law

The first law of thermodynamics may be stated thus:

The increase in internal energy of a body is equal to the heat supplied to the body minus work done by the body

For a thermodynamic cycle, the heat supplied to the system, minus that removed from it, equals the net work done by the system.

More specifically, the First Law encompasses several principles:

- The law of conservation of energy.

  This states that energy can be neither created nor destroyed. However, energy can change forms, and energy can flow from one place to another. The total energy of an isolated system remains the same.

- The concept of internal energy and its relationship to temperature.

  If a system has a definite temperature, then its total energy has three distinguishable components. If it is in motion, it has kinetic energy. If it is in a field (e.g. gravity), it has potential energy. And it has internal energy which is the sum of the kinetic energy of microscopic motions of its constituent atoms, and of the potential energy of interactions between them. Other things being equal, the internal kinetic energy increases as the system's temperature increases. The concept of internal energy is the characteristic distinguishing feature of the first law of thermodynamics.

- The flow of heat is a form of energy transfer.

  In other words, a quantity of heat that flows from a hot body to a cold one can be expressed as an amount of energy being transferred from the hot body to the cold one.

- Performing work is a form of energy transfer.

  For example, when a machine lifts a heavy object upwards, some energy is transferred from the machine to the object. The object acquires its energy in the form of gravitational potential energy in this example.

Combining these principles leads to one traditional statement of the first law of thermodynamics: it is not possible to construct a perpetual motion machine which will continuously do work without consuming energy.

Second law

The second law of thermodynamics asserts the existence of a quantity called the entropy of a system and further states that

When two initially isolated systems in separate but nearby regions of space, each in thermodynamic equilibrium in itself but not necessarily with each other, are then allowed to interact, they will eventually reach a mutual thermodynamic equilibrium. The sum of the entropies of the initially isolated systems is less than the entropy of the final combination. In other words, in the process of reaching a new thermodynamic equilibrium, total entropy increased.

It follows that the entropy of an isolated macroscopic system never decreases. The second law states that spontaneous natural processes increase entropy overall, or in another formulation that heat can spontaneously be conducted or radiated only from a higher-temperature region to a lower-temperature region, but not the other way around.

The second law refers to a wide variety of processes, reversible and irreversible. Its main import is to tell about irreversibility.

The prime example of irreversibility is in the transfer of heat by conduction or radiation. It was known long before the discovery of the notion of entropy that when two bodies of different temperatures, then heat always flows from the hotter body to the colder one.

The second law tells also about kinds of irreversibility other than heat transfer, and the notion of entropy is needed to provide that wider scope of the law.

According to the second law of thermodynamics, in a reversible heat transfer, an element of heat transferred, äQ, is the product of the temperature (T), both of the system and of the sources or destination of the heat, with the increment (dS) of the system's conjugate variable, its entropy (S)

Entropy may also be viewed as a measure of the lack of physical information about the microscopic details of the motion and configuration of a system. The law asserts that for two given macroscopically specified states of a system, there is a quantity called the difference of entropy between them. This entropy difference defines how much additional microscopic physical information is needed to specify one of the macroscopically specified states, given the macroscopic specification of the other - often a conveniently chosen reference state which may be presupposed to exist rather than explicitly stated. A final condition of a natural process always contains microscopically specifiable effects which are not fully and exactly predictable from the macroscopic specification of the initial condition of the process. This is why entropy increases in natural processes - the increase tells how much extra microscopic information is needed to distinguish the final macroscopically specified state from the initial macroscopically specified state.

Third law

The third law of thermodynamics is sometimes stated as follows: The entropy of a perfect crystal at absolute zero is exactly equal to zero.

At zero temperature the system must be in a state with the minimum thermal energy. This statement holds true if the perfect crystal has only one state with minimum energy. Entropy is related to the number of possible microstates according to S = kBln(Ù), where S is the entropy of the system, kB Boltzmann's constant, and Ù the number of microstates (e.g. possible configurations of atoms). At absolute zero there is only 1 microstate possible (Ù=1) and ln(1) = 0.

A more general form of the third law that applies to systems such as glasses that may have more than one minimum energy state:

The entropy of a system approaches a constant value as the temperature approaches zero.

The constant value (not necessarily zero) is called the residual entropy of the system.

Potential vs. Kinetic energy | Back to Top

Potential energy, as the name implies, is energy that has not yet been used, thus the term potential. Kinetic energy is energy in use (or motion). A tank of gasoline has a certain potential energy that is converted into kinetic energy by the engine. When the potential is used up, you're outta gas! Batteries, when new or recharged, have a certain potential. When placed into a tape recorder and played at loud volume (the only settings for such things), the potential in the batteries is transformed into kinetic energy to drive the speakers. When the potential energy is all used up, the batteries are dead. In the case of rechargeable batteries, their potential is reelevated or restored.

In the hydrologic cycle, the sun is the ultimate source of energy, evaporating water (in a fashion raising it's potential above water in the ocean). When the water falls as rain (or snow) it begins to run downhill toward sea-level. As the water get closer to sea-level, it's potential energy is decreased. Without the sun, the water would eventually still reach sea-level, but never be evaporated to recharge the cycle.

Chemicals may also be considered from a potential energy or kinetic energy standpoint. One pound of sugar has a certain potential energy. If that pound of sugar is burned the energy is released all at once. The energy released is kinetic energy (heat). So much is released that organisms would burn up if all the energy was released at once. Organisms must release the energy a little bit at a time. Energy is defined as the ability to do work. Cells convert potential energy, usually in the from of C-C covalent bonds or ATP molecules, into kinetic energy to accomplish cell division, growth, biosynthesis, and active transport, among other things.

BRANCHES OF DESCRIPTION

Thermodynamic systems are theoretical constructions used to model physical systems that exchange matter and energy in terms of the laws of thermodynamics. The study of thermodynamical systems has developed into several related branches, each using a different fundamental model as a theoretical or experimental basis, or applying the principles to varying types of systems.

Classical thermodynamics

Classical thermodynamics accounts for the adventures of thermodynamic systems in terms, either of their time-invariant equilibrium states, or else of their continually repeated cyclic processes, but, formally, not both in the same account. It uses only time-invariant, or equilibrium, macroscopic quantities measurable in the laboratory, counting as time-invariant a long-term time-average of a quantity, such as a flow, generated by a continually repetitive process. Classical thermodynamics does not admit change over time as a fundamental factor in its account of processes. An equilibrium state stands endlessly without change over time, while a continually repeated cyclic process runs endlessly without change over time.

In the account in terms of equilibrium states of a system, a state of thermodynamic equilibrium in a simple system (as defined below in this article), with no externally imposed force field, is spatially homogeneous.

In the classical account strictly and purely in terms of cyclic processes, the spatial interior of the 'working body' of a cyclic process is not considered; the 'working body' thus does not have a defined internal thermodynamic state of its own because no assumption is made that it should be in thermodynamic equilibrium; only its inputs and outputs of energy as heat and work are considered. It is of course possible, and indeed common, for the account in terms of equilibrium states of a system to describe cycles composed of indefinitely many equilibrium states.

Classical thermodynamics was originally concerned with the transformation of energy in cyclic processes, and the exchange of energy between closed systems defined only by their equilibrium states. For these, the distinction between transfers of energy as heat and as work was central.

As classical thermodynamics developed, the distinction between heat and work became less central. This was because there was more interest in open systems, for which the distinction between heat and work is not simple, and is beyond the scope of the present article. Alongside amount of heat transferred as a fundamental quantity, entropy, considered below, was gradually found to be a more generally applicable concept, especially when chemical reactions are of interest. Massieu in 1869 considered entropy as the basic dependent thermodynamic variable, with energy potentials and the reciprocal of thermodynamic temperature as fundamental independent variables. Massieu functions can be useful in present-day non-equilibrium thermodynamics. In 1875, in the work of Josiah Willard Gibbs, the basic thermodynamic quantities were energy potentials, such as internal energy, as dependent variables, and entropy, considered as a fundamental independent variable.

All actual physical processes are to some degree irreversible. Classical thermodynamics can consider irreversible processes, but its account in exact terms is restricted to variables that refer only to initial and final states of thermodynamic equilibrium, or to rates of input and output that do not change with time. For example, classical thermodynamics can consider long-time-average rates of flows generated by continually repeated irreversible cyclic processes. Also it can consider irreversible changes between equilibrium states of systems consisting of several phases (as defined below in this article), or with removable or replaceable partitions. But for systems that are described in terms of equilibrium states, it considers neither flows, nor spatial inhomogeneities in simple systems with no externally imposed force field such as gravity. In the account in terms of equilibrium states of a system, descriptions of irreversible processes refer only to initial and final static equilibrium states; rates of progress are not considered.

Local equilibrium thermodynamics

Local equilibrium thermodynamics is concerned with the time courses and rates of progress of irreversible processes in systems that are smoothly spatially inhomogeneous. It admits time as a fundamental quantity, but only in a restricted way. Rather than considering time-invariant flows as long-term-average rates of cyclic processes, local equilibrium thermodynamics considers time-varying flows in systems that are described by states of local thermodynamic equilibrium, as follows.

For processes that involve only suitably small and smooth spatial inhomogeneities and suitably small changes with time, a good approximation can be found through the assumption of local thermodynamic equilibrium. Within the large or global region of a process, for a suitably small local region, this approximation assumes that a quantity known as the entropy of the small local region can be defined in a particular way. That particular way of definition of entropy is largely beyond the scope of the present article, but here it may be said that it is entirely derived from the concepts of classical thermodynamics; in particular, neither flow rates nor changes over time are admitted into the definition of the entropy of the small local region. It is assumed without proof that the instantaneous global entropy of a non-equilibrium system can be found by adding up the simultaneous instantaneous entropies of its constituent small local regions. Local equilibrium thermodynamics considers processes that involve the time-dependent production of entropy by dissipative processes, in which kinetic energy of bulk flow and chemical potential energy are converted into internal energy at time-rates that are explicitly accounted for. Time-varying bulk flows and specific diffusional flows are considered, but they are required to be dependent variables, derived only from material properties described only by static macroscopic equilibrium states of small local regions. The independent state variables of a small local region are only those of classical thermodynamics.

Generalized or extended thermodynamics

Like local equilibrium thermodynamics, generalized or extended thermodynamics also is concerned with the time courses and rates of progress of irreversible processes in systems that are smoothly spatially inhomogeneous. It describes time-varying flows in terms of states of suitably small local regions within a global region that is smoothly spatially inhomogeneous, rather than considering flows as time-invariant long-term-average rates of cyclic processes. In its accounts of processes, generalized or extended thermodynamics admits time as a fundamental quantity in a more far-reaching way than does local equilibrium thermodynamics. The states of small local regions are defined by macroscopic quantities that are explicitly allowed to vary with time, including time-varying flows. Generalized thermodynamics might tackle such problems as ultrasound or shock waves, in which there are strong spatial inhomogeneities and changes in time fast enough to outpace a tendency towards local thermodynamic equilibrium. Generalized or extended thermodynamics is a diverse and developing project, rather than a more or less completed subject such as is classical thermodynamics.

For generalized or extended thermodynamics, the definition of the quantity known as the entropy of a small local region is in terms beyond those of classical thermodynamics; in particular, flow rates are admitted into the definition of the entropy of a small local region. The independent state variables of a small local region include flow rates, which are not admitted as independent variables for the small local regions of local equilibrium thermodynamics.

Outside the range of classical thermodynamics, the definition of the entropy of a small local region is no simple matter. For a thermodynamic account of a process in terms of the entropies of small local regions, the definition of entropy should be such as to ensure that the second law of thermodynamics applies in each small local region. It is often assumed without proof that the instantaneous global entropy of a non-equilibrium system can be found by adding up the simultaneous instantaneous entropies of its constituent small local regions. For a given physical process, the selection of suitable independent local non-equilibrium macroscopic state variables for the construction of a thermodynamic description calls for qualitative physical understanding, rather than being a simply mathematical problem concerned with a uniquely determined thermodynamic description. A suitable definition of the entropy of a small local region depends on the physically insightful and judicious selection of the independent local non-equilibrium macroscopic state variables, and different selections provide different generalized or extended thermodynamical accounts of one and the same given physical process. This is one of the several good reasons for considering entropy as an epistemic physical variable, rather than as a simply material quantity. According to a respected author: “There is no compelling reason to believe that the classical thermodynamic entropy is a measurable property of nonequilibrium phenomena, ...”

Statistical thermodynamics

Statistical thermodynamics, also called statistical mechanics, emerged with the development of atomic and molecular theories in the second half of the 19th century and early 20th century. It provides an explanation of classical thermodynamics. It considers the microscopic interactions between individual particles and their collective motions, in terms of classical or of quantum mechanics. Its explanation is in terms of statistics that rest on the fact the system is composed of several species of particles or collective motions, the members of each species respectively being in some sense all alike.

Thermodynamic equilibrium

Equilibrium thermodynamics studies transformations of matter and energy in systems at or near thermodynamic equilibrium. In thermodynamic equilibrium, a system’s properties are, by definition, unchanging in time. In thermodynamic equilibrium no macroscopic change is occurring or can be triggered; within the system, every microscopic process is balanced by its opposite; this is called the principle of detailed balance. A central aim in equilibrium thermodynamics is: given a system in a well-defined initial state, subject to specified constraints, to calculate what the equilibrium state of the system is.

In theoretical studies, it is often convenient to consider the simplest kind of thermodynamic system. This is defined variously by different authors. For the present article, the following definition is convenient, as abstracted from the definitions of various authors. A region of material with all intensive properties continuous in space and time is called a phase. A simple system is for the present article defined as one that consists of a single phase of a pure chemical substance, with no interior partitions.

Within a simple isolated thermodynamic system in thermodynamic equilibrium, in the absence of externally imposed force fields, all properties of the material of the system are spatially homogeneous. Much of the basic theory of thermodynamics is concerned with homogeneous systems in thermodynamic equilibrium.

Most systems found in nature or considered in engineering are not in thermodynamic equilibrium, exactly considered. They are changing or can be triggered to change over time, and are continuously and discontinuously subject to flux of matter and energy to and from other systems. For example, according to Callen, "in absolute thermodynamic equilibrium all radioactive materials would have decayed completely and nuclear reactions would have transmuted all nuclei to the most stable isotopes. Such processes, which would take cosmic times to complete, generally can be ignored.". Such processes being ignored, many systems in nature are close enough to thermodynamic equilibrium that for many purposes their behaviour can be well approximated by equilibrium calculations.

Quasi-static transfers between simple systems are nearly in thermodynamic equilibrium and are reversible

It very much eases and simplifies theoretical thermodynamical studies to imagine transfers of energy and matter between two simple systems that proceed so slowly that at all times each simple system considered separately is near enough to thermodynamic equilibrium. Such processes are sometimes called quasi-static and are near enough to being reversible.

Natural processes are partly described by tendency towards thermodynamic equilibrium and are irreversible

If not initially in thermodynamic equilibrium, simple isolated thermodynamic systems, as time passes, tend to evolve naturally towards thermodynamic equilibrium.

In the absence of externally imposed force fields, they become homogeneous in all their local properties. Such homogeneity is an important characteristic of a system in thermodynamic equilibrium in the absence of externally imposed force fields.

Many thermodynamic processes can be modeled by compound or composite systems, consisting of several or many contiguous component simple systems, initially not in thermodynamic equilibrium, but allowed to transfer mass and energy between them. Natural thermodynamic processes are described in terms of a tendency towards thermodynamic equilibrium within simple systems and in transfers between contiguous simple systems. Such natural processes are irreversible.

Non-equilibrium thermodynamics

Non-equilibrium thermodynamics is a branch of thermodynamics that deals with systems that are not in thermodynamic equilibrium; it is also called thermodynamics of irreversible processes. Non-equilibrium thermodynamics is concerned with transport processes and with the rates of chemical reactions. Non-equilibrium systems can be in stationary states that are not homogeneous even when there is no externally imposed field of force; in this case, the description of the internal state of the system requires a field theory. One of the methods of dealing with non-equilibrium systems is to introduce so-called 'internal variables'. These are quantities that express the local state of the system, besides the usual local thermodynamic variables; in a sense such variables might be seen as expressing the 'memory' of the materials. Hysteresis may sometimes be described in this way. In contrast to the usual thermodynamic variables, 'internal variables' cannot be controlled by external manipulations. This approach is usually unnecessary for gases and liquids, but may be useful for solids. Many natural systems still today remain beyond the scope of currently known macroscopic thermodynamic methods.

States and processes

There are two fundamental kinds of entity in thermodynamics, states of a system, and processes of a system. This allows three fundamental approaches to thermodynamic reasoning, that in terms of states of thermodynamic equilibrium of a system, and that in terms of time-invariant processes of a system, and that in terms of cyclic processes of a system. The approach through states of thermodynamic equilibrium of a system requires a full account of the state of the system as well as a notion of process from one state to another of a system, but may require only an idealized or partial account of the state of the surroundings of the system or of other systems.

The method of description in terms of states of thermodynamic equilibrium has limitations. For example, processes in a region of turbulent flow, or in a burning gas mixture, or in a Knudsen gas may be beyond "the province of thermodynamics". This problem can sometimes be circumvented through the method of description in terms of cyclic or of time-invariant flow processes. This is part of the reason why the founders of thermodynamics often preferred the cyclic process description.

Approaches through processes of time-invariant flow of a system are used for some studies. Some processes, for example Joule-Thomson expansion, are studied through steady-flow experiments, but can be accounted for by distinguishing the steady bulk flow kinetic energy from the internal energy, and thus can be regarded as within the scope of classical thermodynamics defined in terms of equilibrium states or of cyclic processes. Other flow processes, for example thermoelectric effects, are essentially defined by the presence of differential flows or diffusion so that they cannot be adequately accounted for in terms of equilibrium states or classical cyclic processes.

The notion of a cyclic process does not require a full account of the state of the system, but does require a full account of how the process occasions transfers of matter and energy between the principal system (which is often called the working body) and its surroundings, which must include at least two heat reservoirs at different known and fixed temperatures, one hotter than the principal system and the other colder than it, as well as a reservoir that can receive energy from the system as work and can do work on the system. The reservoirs can alternatively be regarded as auxiliary idealized component systems, alongside the principal system. Thus an account in terms of cyclic processes requires at least four contributory component systems. The independent variables of this account are the amounts of energy that enter and leave the idealized auxiliary systems. In this kind of account, the working body is often regarded as a "black box", and its own state is not specified. In this approach, the notion of a properly numerical scale of empirical temperature is a presupposition of thermodynamics, not a notion constructed by or derived from it.

Account in terms of states of thermodynamic equilibrium

When a system is at thermodynamic equilibrium under a given set of conditions of its surroundings, it is said to be in a definite thermodynamic state, which is fully described by its state variables.

If a system is simple as defined above, and is in thermodynamic equilibrium, and is not subject to an externally imposed force field, such as gravity, electricity, or magnetism, then it is homogeneous, that is say, spatially uniform in all respects.

In a sense, a homogeneous system can be regarded as spatially zero-dimensional, because it has no spatial variation.

If a system in thermodynamic equilibrium is homogeneous, then its state can be described by a few physical variables, which are mostly classifiable as intensive variables and extensive variables.

Examples of extensive thermodynamic variables are total mass and total volume. Examples of intensive thermodynamic variables are temperature, pressure, and chemical concentration; intensive thermodynamic variables are defined at each spatial point and each instant of time in a system. Physical macroscopic variables can be mechanical or thermal. Temperature is a thermal variable; according to Guggenheim, "the most important conception in thermodynamics is temperature."

Intensive variables are defined by the property that if any number of systems, each in its own separate homogeneous thermodynamic equilibrium state, all with the same respective values of all of their intensive variables, regardless of the values of their extensive variables, are laid contiguously with no partition between them, so as to form a new system, then the values of the intensive variables of the new system are the same as those of the separate constituent systems. Such a composite system is in a homogeneous thermodynamic equilibrium. Examples of intensive variables are temperature, chemical concentration, pressure, density of mass, density of internal energy, and, when it can be properly defined, density of entropy.

Extensive variables are defined by the property that if any number of systems, regardless of their possible separate thermodynamic equilibrium or non-equilibrium states or intensive variables, are laid side by side with no partition between them so as to form a new system, then the values of the extensive variables of the new system are the sums of the values of the respective extensive variables of the individual separate constituent systems. Obviously, there is no reason to expect such a composite system to be in a homogeneous thermodynamic equilibrium. Examples of extensive variables are mass, volume, and internal energy. They depend on the total quantity of mass in the system.

Though, when it can be properly defined, density of entropy is an intensive variable, for inhomogeneous systems, entropy itself does not fit into this classification of state variables. The reason is that entropy is a property of a system as a whole, and not necessarily related simply to its constituents separately. It is true that for any number of systems each in its own separate homogeneous thermodynamic equilibrium, all with the same values of intensive variables, removal of the partitions between the separate systems results in a composite homogeneous system in thermodynamic equilibrium, with all the values of its intensive variables the same as those of the constituent systems, and it is reservedly or conditionally true that the entropy of such a restrictively defined composite system is the sum of the entropies of the constituent systems. But if the constituent systems do not satisfy these restrictive conditions, the entropy of a composite system cannot be expected to be the sum of the entropies of the constituent systems, because the entropy is a property of the composite system as a whole. Therefore, though under these restrictive reservations, entropy satisfies some requirements for extensivity defined just above, entropy in general does not fit the above definition of an extensive variable.

Being neither an intensive variable nor an extensive variable according to the above definition, entropy is thus a stand-out variable, because it is a state variable of a system as a whole. A non-equilibrium system can have a very inhomogeneous dynamical structure. This is one reason for distinguishing the study of equilibrium thermodynamics from the study of non-equilibrium thermodynamics.

The physical reason for the existence of extensive variables is the time-invariance of volume in a given inertial reference frame, and the strictly local conservation of mass, momentum, angular momentum, and energy. As noted by Gibbs, entropy is unlike energy and mass, because it is not locally conserved. The stand-out quantity entropy is never conserved in real physical processes; all real physical processes are irreversible. The motion of planets seems reversible on a short time scale (millions of years), but their motion, according to Newton's laws, is mathematically an example of deterministic chaos. Eventually a planet suffers an unpredictable collision with an object from its surroundings, outer space in this case, and consequently its future course is radically unpredictable. Theoretically this can be expressed by saying that every natural process dissipates some information from the predictable part of its activity into the unpredictable part. The predictable part is expressed in the generalized mechanical variables, and the unpredictable part in heat.

Other state variables can be regarded as conditionally 'extensive' subject to reservation as above, but not extensive as defined above. Examples are the Gibbs free energy, the Helmholtz free energy, and the enthalpy. Consequently, just because for some systems under particular conditions of their surroundings such state variables are conditionally conjugate to intensive variables, such conjugacy does not make such state variables extensive as defined above. This is another reason for distinguishing the study of equilibrium thermodynamics from the study of non-equilibrium thermodynamics. In another way of thinking, this explains why heat is to be regarded as a quantity that refers to a process and not to a state of a system.

A system with no internal partitions, and in thermodynamic equilibrium, can be inhomogeneous in the following respect: it can consist of several so-called 'phases', each homogeneous in itself, in immediate contiguity with other phases of the system, but distinguishable by their having various respectively different physical characters, with discontinuity of intensive variables at the boundaries between the phases; a mixture of different chemical species is considered homogeneous for this purpose if it is physically homogeneous. For example, a vessel can contain a system consisting of water vapour overlying liquid water; then there is a vapour phase and a liquid phase, each homogeneous in itself, but still in thermodynamic equilibrium with the other phase. For the immediately present account, systems with multiple phases are not considered, though for many thermodynamic questions, multiphase systems are important.

Equation of state

The macroscopic variables of a thermodynamic system in thermodynamic equilibrium, in which temperature is well defined, can be related to one another through equations of state or characteristic equations. They express the constitutive peculiarities of the material of the system. The equation of state must comply with some thermodynamic constraints, but cannot be derived from the general principles of thermodynamics alone.

Thermodynamic processes between states of thermodynamic equilibrium

A thermodynamic process is defined by changes of state internal to the system of interest, combined with transfers of matter and energy to and from the surroundings of the system or to and from other systems. A system is demarcated from its surroundings or from other systems by partitions that more or less separate them, and may move as a piston to change the volume of the system and thus transfer work.

Dependent and independent variables for a process

A process is described by changes in values of state variables of systems or by quantities of exchange of matter and energy between systems and surroundings. The change must be specified in terms of prescribed variables. The choice of which variables are to be used is made in advance of consideration of the course of the process, and cannot be changed. Certain of the variables chosen in advance are called the independent variables. From changes in independent variables may be derived changes in other variables called dependent variables. For example a process may occur at constant pressure with pressure prescribed as an independent variable, and temperature changed as another independent variable, and then changes in volume are considered as dependent. Careful attention to this principle is necessary in thermodynamics.

Changes of state of a system

In the approach through equilibrium states of the system, a process can be described in two main ways.

In one way, the system is considered to be connected to the surroundings by some kind of more or less separating partition, and allowed to reach equilibrium with the surroundings with that partition in place. Then, while the separative character of the partition is kept unchanged, the conditions of the surroundings are changed, and exert their influence on the system again through the separating partition, or the partition is moved so as to change the volume of the system; and a new equilibrium is reached. For example, a system is allowed to reach equilibrium with a heat bath at one temperature; then the temperature of the heat bath is changed and the system is allowed to reach a new equilibrium; if the partition allows conduction of heat, the new equilibrium is different from the old equilibrium.

In the other way, several systems are connected to one another by various kinds of more or less separating partitions, and to reach equilibrium with each other, with those partitions in place. In this way, one may speak of a 'compound system'. Then one or more partitions is removed or changed in its separative properties or moved, and a new equilibrium is reached. The Joule-Thomson experiment is an example of this; a tube of gas is separated from another tube by a porous partition; the volume available in each of the tubes is determined by respective pistons; equilibrium is established with an initial set of volumes; the volumes are changed and a new equilibrium is established. Another example is in separation and mixing of gases, with use of chemically semi-permeable membranes.

Commonly considered thermodynamic processes

It is often convenient to study a thermodynamic process in which a single variable, such as temperature, pressure, or volume, etc., is held fixed. Furthermore, it is useful to group these processes into pairs, in which each variable held constant is one member of a conjugate pair.

Several commonly studied thermodynamic processes are:

- Isobaric process: occurs at constant pressure
- Isochoric process: occurs at constant volume (also called isometric/isovolumetric)
- Isothermal process: occurs at a constant temperature
- Adiabatic process: occurs without loss or gain of energy as heat
- Isentropic process: a reversible adiabatic process occurs at a constant entropy, but is a fictional idealization. Conceptually it is possible to actually physically conduct a process that keeps the entropy of the system constant, allowing systematically controlled removal of heat, by conduction to a cooler body, to compensate for entropy produced within the system by irreversible work done on the system. Such isentropic conduct of a process seems called for when the entropy of the system is considered as an independent variable, as for example when the internal energy is considered as a function of the entropy and volume of the system, the natural variables of the internal energy as studied by Gibbs.
- Isenthalpic process: occurs at a constant enthalpy
- Isolated process: no matter or energy (neither as work nor as heat) is transferred into or out of the system

It is sometimes of interest to study a process in which several variables are controlled, subject to some specified constraint. In a system in which a chemical reaction can occur, for example, in which the pressure and temperature can affect the equilibrium composition, a process might occur in which temperature is held constant but pressure is slowly altered, just so that chemical equilibrium is maintained all the way. There is a corresponding process at constant temperature in which the final pressure is the same but is reached by a rapid jump. Then it can be shown that the volume change resulting from the rapid jump process is smaller than that from the slow equilibrium process. The work transferred differs between the two processes.

ACCOUNT IN TERMS OF CYCLIC PROCESSES

A cyclic process is a process that can be repeated indefinitely often without changing the final state of the system in which the process occurs. The only traces of the effects of a cyclic process are to be found in the surroundings of the system or in other systems. This is the kind of process that concerned early thermodynamicists such as Carnot, and in terms of which Kelvin defined absolute temperature, before the use of the quantity of entropy by Rankine and its clear identification by Clausius. For some systems, for example with some plastic working substances, cyclic processes are practically nearly unfeasible because the working substance undergoes practically irreversible changes. This is why mechanical devices are lubricated with oil and one of the reasons why electrical devices are often useful.

A cyclic process of a system requires in its surroundings at least two heat reservoirs at different temperatures, one at a higher temperature that supplies heat to the system, the other at a lower temperature that accepts heat from the system. The early work on thermodynamics tended to use the cyclic process approach, because it was interested in machines that converted some of the heat from the surroundings into mechanical power delivered to the surroundings, without too much concern about the internal workings of the machine. Such a machine, while receiving an amount of heat from a higher temperature reservoir, always needs a lower temperature reservoir that accepts some lesser amount of heat, the difference in amounts of heat being converted to work. Later, the internal workings of a system became of interest, and they are described by the states of the system. Nowadays, instead of arguing in terms of cyclic processes, some writers are inclined to derive the concept of absolute temperature from the concept of entropy, a variable of state.

Instrumentation

There are two types of thermodynamic instruments, the meter and the reservoir. A thermodynamic meter is any device that measures any parameter of a thermodynamic system. In some cases, the thermodynamic parameter is actually defined in terms of an idealized measuring instrument. For example, the zeroth law states that if two bodies are in thermal equilibrium with a third body, they are also in thermal equilibrium with each other. This principle, as noted by James Maxwell in 1872, asserts that it is possible to measure temperature. An idealized thermometer is a sample of an ideal gas at constant pressure. From the ideal gas law PV=nRT, the volume of such a sample can be used as an indicator of temperature; in this manner it defines temperature. Although pressure is defined mechanically, a pressure-measuring device, called a barometer may also be constructed from a sample of an ideal gas held at a constant temperature. A calorimeter is a device that measures and define the internal energy of a system.

A thermodynamic reservoir is a system so large that it does not appreciably alter its state parameters when brought into contact with the test system. It is used to impose a particular value of a state parameter upon the system. For example, a pressure reservoir is a system at a particular pressure, which imposes that pressure upon any test system that it is mechanically connected to. The Earth's atmosphere is often used as a pressure reservoir.

Scope of thermodynamics

Originally thermodynamics concerned material and radiative phenomena that are experimentally reproducible. For example, a state of thermodynamic equilibrium is a steady state reached after a system has aged so that it no longer changes with the passage of time. But more than that, for thermodynamics, a system, defined by its being prepared in a certain way must, consequent on every particular occasion of preparation, upon aging, reach one and the same eventual state of thermodynamic equilibrium, entirely determined by the way of preparation. Such reproducibility is because the systems consist of so many molecules that the molecular variations between particular occasions of preparation have negligible or scarcely discernable effects on the macroscopic variables that are used in thermodynamic descriptions. This led to Boltzmann's discovery that entropy had a statistical or probabilistic nature. Probabilistic and statistical explanations arise from the experimental reproducibility of the phenomena.

Gradually, the laws of thermodynamics came to be used to explain phenomena that occur outside the experimental laboratory. For example, phenomena on the scale of the earth's atmosphere cannot be reproduced in a laboratory experiment. But processes in the atmosphere can be modeled by use of thermodynamic ideas, extended well beyond the scope of laboratory equilibrium thermodynamics. A parcel of air can, near enough for many studies, be considered as a closed thermodynamic system, one that is allowed to move over significant distances. The pressure exerted by the surrounding air on the lower face of a parcel of air may differ from that on its upper face. If this results in rising of the parcel of air, it can be considered to have gained potential energy as a result of work being done on it by the combined surrounding air below and above it. As it rises, such a parcel usually expands because the pressure is lower at the higher altitudes that it reaches. In that way, the rising parcel also does work on the surrounding atmosphere. For many studies, such a parcel can be considered nearly to neither gain nor lose energy by heat conduction to its surrounding atmosphere, and its rise is rapid enough to leave negligible time for it to gain or lose heat by radiation; consequently the rising of the parcel is near enough adiabatic. Thus the adiabatic gas law accounts for its internal state variables, provided that there is no precipitation into water droplets, no evaporation of water droplets, and no sublimation in the process. More precisely, the rising of the parcel is likely to occasion friction and turbulence, so that some potential and some kinetic energy of bulk converts into internal energy of air considered as effectively stationary. Friction and turbulence thus oppose the rising of the parcel.

Chemical reaction

A chemical reaction is a process that leads to the transformation of one set of chemical substances to another. Classically, chemical reactions encompass changes that strictly involve the motion of electrons in the forming and breaking of chemical bonds between atoms, and can often be described by a chemical equation. The substance (or substances) initially involved in a chemical reaction are called reactants or reagents. Chemical reactions are usually characterized by a chemical change, and they yield one or more products, which usually have properties different from the reactants. Reactions often consist of a sequence of individual sub-steps, the so-called elementary reactions, and the information on the precise course of action is part of the reaction mechanism. Chemical reactions are described with chemical equations, which graphically present the starting materials, end products, and sometimes intermediate products and reaction conditions.

Chemical reactions happen at a characteristic reaction rate at a given temperature and chemical concentration, and rapid reactions are often described as spontaneous, requiring no input of extra energy other than thermal energy. Non-spontaneous reactions run so slowly that they are considered to require the input of some type of additional energy (such as extra heat, light or electricity) in order to proceed to completion (chemical equilibrium) at human time scales.

Different chemical reactions are used in combinations during in chemical synthesis in order to obtain a desired product. In biochemistry, a similar series of chemical reactions form metabolic pathways. These reactions are often catalyzed by protein enzymes. These enzymes increase the rates of biochemical reactions, so that metabolic syntheses and decompositions impossible under ordinary conditions may be performed at the temperatures and concentrations present within a cell.

The general concept of a chemical reaction has been extended to non-chemical reactions between entities smaller than atoms, including nuclear reactions, radioactive decays, and reactions between elementary particles as described by quantum field theory.

In most cases of interest in chemical thermodynamics there are internal degrees of freedom and processes, such as chemical reactions and phase transitions, which always create entropy unless they are at equilibrium, or are maintained at a "running equilibrium" through "quasi-static" changes by being coupled to constraining devices, such as pistons or electrodes, to deliver and receive external work. Even for homogeneous "bulk" materials, the free energy functions depend on the composition, as do all the extensive thermodynamic potentials, including the internal energy. If the quantities { Ni }, the number of chemical species, are omitted from the formulae, it is impossible to describe compositional changes.

Fundamental concepts

Thermodynamic states

The application of thermodynamic principles begins by defining a system that is in some sense distinct from its surroundings. For example, the system could be a sample of gas inside a cylinder with a movable piston, an entire steam engine, a marathon runner, the planet Earth, a neutron star, a black hole, or even the entire universe. In general, systems are free to exchange heat, work, and other forms of energy with their surroundings.

A system's condition at any given time is called its thermodynamic state. For a gas in a cylinder with a movable piston, the state of the system is identified by the temperature, pressure, and volume of the gas. These properties are characteristic parameters that have definite values at each state and are independent of the way in which the system arrived at that state. In other words, any change in value of a property depends only on the initial and final states of the system, not on the path followed by the system from one state to another. Such properties are called state functions. In contrast, the work done as the piston moves and the gas expands and the heat the gas absorbs from its surroundings depend on the detailed way in which the expansion occurs.

The behaviour of a complex thermodynamic system, such as Earth's atmosphere, can be understood by first applying the principles of states and properties to its component parts—in this case, water, water vapour, and the various gases making up the atmosphere. By isolating samples of material whose states and properties can be controlled and manipulated, properties and their interrelations can be studied as the system changes from state to state.

Thermodynamic equilibrium

A particularly important concept is thermodynamic equilibrium, in which there is no tendency for the state of a system to change spontaneously. For example, the gas in a cylinder with a movable piston will be at equilibrium if the temperature and pressure inside are uniform and if the restraining force on the piston is just sufficient to keep it from moving. The system can then be made to change to a new state only by an externally imposed change in one of the state functions, such as the temperature by adding heat or the volume by moving the piston. A sequence of one or more such steps connecting different states of the system is called a process. In general, a system is not in equilibrium as it adjusts to an abrupt change in its environment. For example, when a balloon bursts, the compressed gas inside is suddenly far from equilibrium, and it rapidly expands until it reaches a new equilibrium state. However, the same final state could be achieved by placing the same compressed gas in a cylinder with a movable piston and applying a sequence of many small increments in volume (and temperature), with the system being given time to come to equilibrium after each small increment. Such a process is said to be reversible because the system is at (or near) equilibrium at each step along its path, and the direction of change could be reversed at any point. This example illustrates how two different paths can connect the same initial and final states. The first is irreversible (the balloon bursts), and the second is reversible. The concept of reversible processes is something like motion without friction in mechanics. It represents an idealized limiting case that is very useful in discussing the properties of real systems. Many of the results of thermodynamics are derived from the properties of reversible processes.

Temperature

The concept of temperature is fundamental to any discussion of thermodynamics, but its precise definition is not a simple matter. For example, a steel rod feels colder than a wooden rod at room temperature simply because steel is better at conducting heat away from the skin. It is therefore necessary to have an objective way of measuring temperature. In general, when two objects are brought into thermal contact, heat will flow between them until they come into equilibrium with each other. When the flow of heat stops, they are said to be at the same temperature. The zeroth law of thermodynamics formalizes this by asserting that if an object A is in simultaneous thermal equilibrium with two other objects B and C, then B and C will be in thermal equilibrium with each other if brought into thermal contact. Object A can then play the role of a thermometer through some change in its physical properties with temperature, such as its volume or its electrical resistance.

With the definition of equality of temperature in hand, it is possible to establish a temperature scale by assigning numerical values to certain easily reproducible fixed points. For example, in the Celsius (°C) temperature scale, the freezing point of pure water is arbitrarily assigned a temperature of 0 °C and the boiling point of water the value of 100 °C (in both cases at 1 standard atmosphere; see atmospheric pressure). In the Fahrenheit (°F) temperature scale, these same two points are assigned the values 32 °F and 212 °F, respectively. There are absolute temperature scales related to the second law of thermodynamics. The absolute scale related to the Celsius scale is called the Kelvin (K) scale, and that related to the Fahrenheit scale is called the Rankine (°R) scale. These scales are related by the equations K = °C + 273.15, °R = °F + 459.67, and °R = 1.8 K. Zero in both the Kelvin and Rankine scales is at absolute zero.

Work and energy

Energy has a precise meaning in physics that does not always correspond to everyday language, and yet a precise definition is somewhat elusive. The word is derived from the Greek word ergon, meaning work, but the term work itself acquired a technical meaning with the advent of Newtonian mechanics. For example, a man pushing on a car may feel that he is doing a lot of work, but no work is actually done unless the car moves. The work done is then the product of the force applied by the man multiplied by the distance through which the car moves. If there is no friction and the surface is level, then the car, once set in motion, will continue rolling indefinitely with constant speed. The rolling car has something that a stationary car does not have—it has kinetic energy of motion equal to the work required to achieve that state of motion. The introduction of the concept of energy in this way is of great value in mechanics because, in the absence of friction, energy is never lost from the system, although it can be converted from one form to another. For example, if a coasting car comes to a hill, it will roll some distance up the hill before coming to a temporary stop. At that moment its kinetic energy of motion has been converted into its potential energy of position, which is equal to the work required to lift the car through the same vertical distance. After coming to a stop, the car will then begin rolling back down the hill until it has completely recovered its kinetic energy of motion at the bottom. In the absence of friction, such systems are said to be conservative because at any given moment the total amount of energy (kinetic plus potential) remains equal to the initial work done to set the system in motion.

As the science of physics expanded to cover an ever-wider range of phenomena, it became necessary to include additional forms of energy in order to keep the total amount of energy constant for all closed systems (or to account for changes in total energy for open systems). For example, if work is done to accelerate charged particles, then some of the resultant energy will be stored in the form of electromagnetic fields and carried away from the system as radiation. In turn the electromagnetic energy can be picked up by a remote receiver (antenna) and converted back into an equivalent amount of work. With his theory of special relativity, Albert Einstein realized that energy (E) can also be stored as mass (m) and converted back into energy, as expressed by his famous equation E = mc2, where c is the velocity of light. All of these systems are said to be conservative in the sense that energy can be freely converted from one form to another without limit. Each fundamental advance of physics into new realms has involved a similar extension to the list of the different forms of energy. In addition to preserving the first law of thermodynamics (see below), also called the law of conservation of energy, each form of energy can be related back to an equivalent amount of work required to set the system into motion.

Thermodynamics encompasses all of these forms of energy, with the further addition of heat to the list of different kinds of energy. However, heat is fundamentally different from the others in that the conversion of work (or other forms of energy) into heat is not completely reversible, even in principle. In the example of the rolling car, some of the work done to set the car in motion is inevitably lost as heat due to friction, and the car eventually comes to a stop on a level surface. Even if all the generated heat were collected and stored in some fashion, it could never be converted entirely back into mechanical energy of motion. This fundamental limitation is expressed quantitatively by the second law of thermodynamics (see below).

The role of friction in degrading the energy of mechanical systems may seem simple and obvious, but the quantitative connection between heat and work, as first discovered by Count Rumford, played a key role in understanding the operation of steam engines in the 19th century and similarly for all energy-conversion processes today.

Total internal energy

Although classical thermodynamics deals exclusively with the macroscopic properties of materials—such as temperature, pressure, and volume—thermal energy from the addition of heat can be understood at the microscopic level as an increase in the kinetic energy of motion of the molecules making up a substance. For example, gas molecules have translational kinetic energy that is proportional to the temperature of the gas: the molecules can rotate about their centre of mass, and the constituent atoms can vibrate with respect to each other (like masses connected by springs). Additionally, chemical energy is stored in the bonds holding the molecules together, and weaker long-range interactions between the molecules involve yet more energy. The sum total of all these forms of energy constitutes the total internal energy of the substance in a given thermodynamic state. The total energy of a system includes its internal energy plus any other forms of energy, such as kinetic energy due to motion of the system as a whole (e.g., water flowing through a pipe) and gravitational potential energy due to its elevation.

BIBLIOGRAPHY

- Alternative bioseparation operations: life beyond packed-bed chromatography T.M. Przybycien, N.S. Pujar and L.M. Steele Curr Opin Biotechnol, 15 (5) (2004), pp. 469-478
- Bard, A.J.; Faulkner, L.R. Electrochemical Methods: Fundamentals and Applications. New York: John Wiley & Sons, 2nd Edition, 2000.
- Bernard F (2003). Handbook of Thin-Layer Chromatography. Marcel Dekker Inc. ISBN 978-0824748661. OCLC 437068122.
- Borman, Stu (1987). Eluent, Effluent, Eluate, and Eluite. Analytical Chemistry. 59 (2): 99A. doi:10.1021/ac00129a735.
- Bourgeois S, Pfahl M (1976). Repressors. In Anfinsen CB, Edsall JT, Richards FM (eds.). Advances in Protein Chemistry. Vol. 30. pp. 6–7. doi:10.1016/S0065-3233(08)60478-7. ISBN 978-0-12-034230-3. PMID 779429.
- Displacement Chromatography 101 Archived 15 September 2008 at the Wayback Machine. Sachem, Inc. Austin, TX 78737
- Dr. Irene Novaczek. Canada's Fossil Fuel Dependency. Elements. Retrieved 18 January 2007.
- E.A. Guggenheim (1967). Thermodynamics: An Advanced Treatment for Chemists and Physicists, 5th ed. North Holland; John Wiley & Sons (Interscience). A remarkably astute treatise. Library of Congress Catalog No. 67-20003
- Elert, Glenn (2007). Chemical Potential Energy. The Physics Hypertextbook. Retrieved 11 September 2007.
- Ellis LA, Roberts DJ (1997). Chromatographic and hyphenated methods for elemental speciation analysis in environmental media. Journal of Chromatography A. 774 (1–2): 3–19. doi:10.1016/S0021-9673(97)00325-7. PMID 9253184.

- Ettre LS (1993). Nomenclature for chromatography (IUPAC Recommendations 1993). Pure and Applied Chemistry. 65 (4): 819–872. doi:10.1351/pac199365040819.
- Ettre LS, Sakodynskii KI (March 1993). M. S. Tswett and the discovery of chromatography II: Completion of the development of chromatography (1903–1910). Chromatographia. 35 (5–6): 329–338. doi:10.1007/BF02277520. S2CID 97052560.
- Ettre LS, Zlatkis A, eds. (26 August 2011). 75 Years of Chromatography: A Historical Dialogue. Elsevier. ISBN 978-0-08-085817-3.
- Fewell, M. P. (1995). The atomic nuclide with the highest mean binding energy. American Journal of Physics. 63 (7): 653–658. Bibcode:1995AmJPh..63..653F. doi:10.1119/1.17828.
- Fossil fuel. Archived from the original on 10 May 2012.
- Fossil fuel. EPA. Archived from the original on 12 March 2007. Retrieved 18 January 2007.
- González-González, Mirna; Mayolo-Deloisa, Karla; Rito-Palomares, Marco (1 January 2020), Matte, Allan (ed.), Chapter 5 - Recent advances in antibody-based monolith chromatography for therapeutic applications, Approaches to the Purification, Analysis and Characterization of Antibody-Based Therapeutics, Elsevier, pp. 105–116, doi:10.1016/b978-0-08-103019-6.00005-9, ISBN 978-0-08-103019-6, S2CID 226450210, retrieved 31 January 2022
- Guetens G, De Boeck G, Highley MS, Wood M, Maes RA, Eggermont AA, Hanauske A, de Bruijn EA, Tjaden UR (2002). Hyphenated techniques in anticancer drug monitoring. II. Liquid chromatography-mass spectrometry and capillary electrophoresis-mass spectrometry. Journal of Chromatography A. 976 (1–2): 239–47. doi:10.1016/S0021-9673(02)01227-X. PMID 12462615.
- Guetens G, De Boeck G, Wood M, Maes RA, Eggermont AA, Highley MS, van Oosterom AT, de Bruijn EA, Tjaden UR (2002). Hyphenated techniques in anticancer drug monitoring. I. Capillary gas chromatography-mass spectrometry. Journal of Chromatography A. 976 (1–2): 229–38. doi:10.1016/S0021-9673(02)01228-1. PMID 12462614.
- Harper, Douglas. chromatography. Online Etymology Dictionary.

- Harwood LM, Moody CJ (1989). Experimental organic chemistry: Principles and Practice (Illustrated ed.). WileyBlackwell. pp. 180–185. ISBN 978-0-632-02017-1.
- Herbert B. Callen (1960). Thermodynamics. Wiley & Sons. The clearest account of the logical foundations of the subject. ISBN 0-471-13035-4. Library of Congress Catalog No. 60-5597
- Holt RM, Newman MJ, Pullen FS, Richards DS, Swanson AG (1997). High-performance liquid chromatography/NMR spectrometry/mass spectrometry: further advances in hyphenated technology. Journal of Mass Spectrometry. 32 (1): 64–70. Bibcode:1997JMSp...32...64H. doi:10.1002/(SICI)1096-9888(199701)32:1<64::AID-JMS450>3.0.CO;2-7. PMID 9008869.
- Hostettmann K, Marston A, Hostettmann M (1998). Preparative Chromatography Techniques Applications in Natural Product Isolation (Second ed.). Berlin, Heidelberg: Springer Berlin Heidelberg. p. 50. ISBN 9783662036310.
- Ilya Prigogine & R. Defay, translated by D.H. Everett; Chapter IV (1954). Chemical Thermodynamics. Longmans, Green & Co. Exceptionally clear on the logical foundations as applied to chemistry; includes non-equilibrium thermodynamics.
- Ilya Prigogine (1967). Thermodynamics of Irreversible Processes, 3rd ed. Interscience: John Wiley & Sons. A simple, concise monograph explaining all the basic ideas. Library of Congress Catalog No. 67-29540
- International Energy Annual 2006. Archived from the original on 5 February 2009. Retrieved 8 February 2009.
- Manish T. How does column chromatography work?. BrightMags. Archived from the original on 21 April 2017. Retrieved 7 April 2017.
- Ongkudon, Clarence M.; Kansil, Tamar; Wong, Charlotte (2014). Challenges and strategies in the preparation of large-volume polymer-based monolithic chromatography adsorbents. Journal of Separation Science. 37 (5): 455–464. doi:10.1002/jssc.201300995. ISSN 1615-9314. PMID 24376196.
- Rahman, M.; El-Aty, A.M.A.; Choi, J.-H.; Shin, H.-C.; Shin, S.-C.; Shim, J.-H. (November 2015). Chapter 3 Basic Overview on Gas Chromatography Columns. Analytical Separation Science. pp. 823–834. ISBN 9783527333745.

- Rincon, Paul (22 March 2004). Bones hint at first use of fire. BBC News. Retrieved 11 September 2007.
- Singh NK, DSouza RN, Bibi NS, Fernández-Lahore M (2015). Direct Capture of His6-Tagged Proteins Using Megaporous Cryogels Developed for Metal-Ion Affinity Chromatography. In Reichelt S (ed.). Affinity Chromatography. Methods in Molecular Biology. Vol. 1286. pp. 201–12. doi:10.1007/978-1-4939-2447-9_16. ISBN 978-1-4939-2447-9. PMID 25749956.
- Skoog, D.A.; West, D.M.; Holler, F.J. Fundamentals of Analytical Chemistry New York: Saunders College Publishing, 5th Edition, 1988.
- Still WC, Kahn M, Mitra A (1978). Rapid chromatographic technique for preparative separations with moderate resolution. J. Org. Chem. 43 (14): 2923–2925. CiteSeerX 10.1.1.476.6501. doi:10.1021/jo00408a041.
- Th. De Donder (1922). L'affinite. Applications aux gaz parfaits. Bulletin de la Classe des Sciences, Académie Royale de Belgique. Series 5. 8: 197–205.
- Th. De Donder (1922). Sur le theoreme de Nernst. Bulletin de la Classe des Sciences, Académie Royale de Belgique. Series 5. 8: 205–210.

INDEX

A

Ability 56, 63-64, 68, 86, 215
Abrasion 65
Absorption 64, 117, 148-155
Acid-base 122-123
Acrylic 52, 72, 210
Additives 6, 74, 186
Adhesion 50, 215
Adiabatic 236, 238
Adsorbent 99
Adsorption 105, 172
Aerospace 71
Aircraft 77
Alcohols 44
Alkalinity 2
Alkane 2
Alloys 7, 60, 92, 187
Alumina 27-28, 31, 37, 49
Aluminium 49, 66, 99
Ammonium 58, 182
Animals 74
Anions 119
Anisotropy 92
Annealing 202
Anthracite 141
Anticancer 246
Artifacts 13-14
Atmosphere 33, 37, 185, 238, 240
Automobile 70
Auxiliary 159, 231

B

Backbone 208
Backside 195
Bacteria 9, 136
Balloon 241
Barium 184
Baseline 100
Benzoyl 209
Berlin 247
Binary 36
Binder 46
Binding 28-30, 128
Biodiesel 130
Biology 126
Biomedical 77
Bromine 116, 131
Burette 121, 124
Burner 138, 140

Bursts 241

Butane 145

Butylene 145

C

Calcium 9, 31-33, 136, 169

Calibrated 175

Capacity 45, 64, 88, 190

Carbide 5, 13, 15, 36

Carbon 5, 7, 22, 75, 104, 116, 135-136, 140-142, 183, 198

Carbonates 9

Carboxylic 130

Casein 52-53

Cathode 166

Cationic 210

Cations 119

Ceilings 35-36

Cellulose 99

Celsius 242

Cements 28, 30

Ceramics 3, 5, 9-16, 18, 76, 197

Changes 39, 124, 128-129, 150, 154, 162, 173-174, 188, 191, 200, 205, 217, 219, 226-227, 234-235,

Charcoal 132

Chemistry 83, 102, 114-115, 120, 190, 245, 247

Complexity 216

Components 11, 18, 20, 52, 93-96, 100-101, 108-109, 149, 201

Composites 19-22, 71, 75-77, 186

Compounds 28, 31, 36, 42, 45, 55, 97-99, 101, 103, 116, 135, 150, 165, 216

Cooking 132

Cooling 73, 86-87, 187, 199, 202

Copolymers 81, 85

Copper 119, 187

Corrosion 1-2, 59

Co-solvent 104

Coulometry 157, 161

Crosslinks 201

Crucible 141-142, 168-169

Cryogenic 42-43

Crystal 3, 27, 163-165, 171, 183-184, 224

Crystals 164-165, 170, 184, 200

Cylinder 56, 240-241

D

Damage 140

Decomposes 44

Density 21, 56-57, 60-63, 76, 87, 137, 188, 232

Deposits 39

Developed 121, 190

Diameters 56

Diamond 47

Dielectric 185, 188-189

Digestion 168, 170-171

Dimensions 38, 152

Dioxide 58, 104, 135, 183

Dipoles 189

Dipped 159

Dispersion 26, 79

Dissolved 94, 131, 177-180, 182-183

Diversity 18

Droplets 79, 83, 238
Dryers 18
Ductile 68

E

Effluent 108-109
Elasticity 66
Elastomers 73, 81
Electrodes 157-158, 162, 212, 214
Electrons 91-92, 149, 151, 156, 158, 162-163, 208
Elements 3-4, 7, 87, 89, 141, 170
Eluent 98
Elution 100, 108
Emission 117, 146-148, 154-155
Endpoint 121, 123, 125, 128-130
Energy 53, 55-57, 87-89, 132-134, 137, 147-151, 154, 195, 211, 213, 217-220, 222, 224-227, 229, 231-234, 236, 238-240, 242-244
Engine 11, 15, 41
Enthalpy 89, 217
Entropy 89, 217, 220, 223-224, 226-228, 232-233, 236-237
Enzymes 239
Ethylene 208, 213
Explosion 137
External 60, 76, 199

F

Fibreboard 36
Fibres 21, 23
Filtering 128
Fluorine 43
Fossil 133-136, 246
Fragments 13, 107
Freezing 43-44, 86
Furnace 4, 38, 140, 142, 144, 196

G

Gaseous 134, 136, 144-145
Geometries 194
Glasses 5, 26, 88, 91
Glassware 27
Glassy 86, 200
Glucose 131
Grains 47-48
Graphite 5
Gravimeter 173-174, 176
Gravimetry 166, 172, 174
Gravity 21, 111-112, 138, 173-176
Growth 8, 170, 210
Gypsum 28, 32, 34-36

H

Hafnium 36
Hardening 28
Hardens 30
Hardness 7, 47, 65, 67
High-tech 16
Homogenous 195
Humans 132
Hybrid 41, 46
Hydraulic 28, 30, 33
Hydrazine 43-44
Hydrogen 42, 44, 135-137, 142, 164, 183
Hydroxide 33, 119
Hydroxyl 131
Hypergolic 43

I

Indicators 122, 126

Instrument 110, 112, 173-174, 176, 185, 190, 192, 194, 197

Internal 35, 171, 215, 222, 230, 232, 234, 236-238, 244

Iodine 125, 131

Iodometric 125

Isothermal 202

Isotope 157

J

Journal 246-247

M

Materials 1-6, 8-10, 12-14, 16-17, 20-23, 25, 28-31, 36-39, 47, 49-53, 58-63, 65-69, 71-72, 74, 76, 79, 85-86, 88, 114-115, 165, 184, 186, 188, 191-196, 198, 206-207

Matrices 24

Matrix 19-25, 48, 87, 97, 100

Metals 1-2, 7, 59-60, 132, 179, 187

Methane 43, 135

Microprobe 156

Microscopy 4, 119, 186

Microwave 149-150

Minerals 8-9, 39, 47-48

Mobile 93-94, 96-97, 101-106, 108-111

N

Neutrons 163

Nitrate 58, 170

Nitrogen 44, 55, 130, 142

O

Orthopedic 12, 16

Oxalate 169

Oxidations 159

Oxides 10, 31, 76

Oxygen 41-43, 45, 59, 131, 136-137, 210

P

Petroleum 42, 134-135, 144

Pharma 186

Polarity 106

Polyatomic 155

Polyester 23-24, 70

Polymers 3, 6, 50, 69-73, 75, 77-81, 84-85, 89, 105, 186, 197, 200, 206-208, 211-212, 214-216

Porcelain 179

Portion 160, 182

Positive 182, 217

Potassium 125-126, 130-131

Precision 173, 179, 196

R

Radiation 117, 147-150, 152-157, 163-164

Reservoirs 231

Residual 140

Residue 140, 180, 183

Resins 24, 52, 74

Resistance 1-2, 5, 32, 35-36, 58, 61-64, 68, 75, 77, 158

Resistant 24, 63

Resolution 175, 192

Resonance 117-118, 149, 193

Resultant 216

Reverse 130, 160

Reversible 73, 202, 223, 229, 241

Revision 90
Robust 20
Rockets 41-42
Rotation 112
Rubbers 73

S

Sea-level 224
Sediment 119, 182
Separation 78, 93, 96, 100, 103-105, 107, 110, 112-114, 119, 247
Signal 95, 118, 128
Silica 5, 26-27, 31, 88, 97
Silicon 5, 15, 36, 76
Silver 119, 167, 170, 184
Sludge 140, 178, 182
Smooth 86
Soda-lime 26
Sodium 27
Softening 62
Softer 47
Solid-like 86
Solution 79-81, 83, 103-105, 108, 119-122, 124-125, 128-130, 151, 157-159, 161-164, 167-168, 170-171
Solvents 39, 96, 110, 186
Strain 66, 193-194, 199, 203-204
Stream 179
Stresses 30, 66, 68, 200
Students 83
Suitable 228
Sunlight 55
Surfactant 79
Surveys 175
Suspension 82-83
Symmetry 163

T

Techniques 4, 18, 79, 83-84, 95, 108-109, 114, 164, 188, 246
Threshold 38, 89

U

Uniformity 213

V

Vacancy 155
Versus 216
Vibrations 90
Viscosity 50, 78, 80-81, 86, 88, 138
Vitamin 131
Volatile 141-142, 178, 182-183
Volumetric 117, 121

W

Wastes 120
Wastewater 178-179
Wavelength 148-149, 152-153, 155
Weathering 39, 64
Window 27

X

X-rays 163

Z

Zirconia 76

9 789390 013876

Printed at Repro India Ltd.